Insights into the New Hydrogen Economy

William J. Nuttall · Joseph B. Powell ·
Karim L. Anaya-Stucchi ·
Adetokunboh T. Bakenne · Andy Wilson

Insights into the New Hydrogen Economy

 Springer

William J. Nuttall
Consultant
Cambridgeshire, UK

Karim L. Anaya-Stucchi
Energy Policy Research Group
University of Cambridge
Cambridge, Cambridgeshire, UK

Andy Wilson
Consultant
Maidenhead, Berkshire, UK

Joseph B. Powell
Houston Energy Transition Institute
University of Houston
Houston, TX, USA

Adetokunboh T. Bakenne
Consultant
London, UK

ISBN 978-3-031-71832-8   ISBN 978-3-031-71833-5 (eBook)
https://doi.org/10.1007/978-3-031-71833-5

This Springer imprint is published by the registered company Springer Nature Switzerland AG
The registered company address is: Gewerbestrasse 11, 6330 Cham, Switzerland

If disposing of this product, please recycle the paper.

Acknowledgements

This project has only been possible thanks to the generous participation of numerous experts from diverse areas of the hydrogen innovation landscape. This work builds upon three expert meetings held in 2021 and 2022. The three events were organised by The Open University as a knowledge exchange activity. The first two events, held on 19 March 2021 and 24 June 2021 were organised virtually in part, as a consequence of the COVID-19 pandemic. The third event was held in person in central London on 4 July 2022. Thanks are due to the panellists from the three events:

- Bernaurdshaw Neppolian
- Naomi Boness
- Mark Crowther
- Bartek Glowacki
- Ian Hibbitt
- Michaela Kendall
- Ian MacGregor
- Margaux Moore
- Joe Powell
- Dan Sadler
- Bo Sears

The three events were structured around expert panels with other delegates also representing expert perspectives. The events were characterised by active dialogue and the process benefitted from interventions by delegates to the meetings. The authors are especially grateful to those delegates who kindly provided insight and comment. Key to the process is the notion that the dialogue is a conversation amongst experts contributing in different ways.

The core ideas of this book were first written up in draft unpublished reports circulated amongst the wider community gathered around this process (delegates and panellists). Written comments were received from the following individuals and the authors are most grateful for their advice and assistance at that stage:

- Bo Sears
- Chris Johnson
- Robin Little
- Margaux Moore

The authors are also most grateful to Luke Christoforidis and Charles Forsberg for helpful conversations.

The ideas and propositions presented in this his book do not necessarily align with the views of the panellists, contributing experts and those that sent in written comments, nor the organisations to which they may be affiliated.

This work also benefits from formal contributions from independent experts. They are the named authors of various sections (text boxes) presented in chapter 5. The project team is most grateful for all those that kindly provided such content and gave permission for its use. Special thanks are due to Joe Powell for agreeing to work with The Open University (OU) on framing and developing this whole activity from the very beginning. The team is also most grateful to the Payne Institute for Public Policy at the Colorado School of Mines, USA, for its assistance with the organisation of the second project webinar. We would also like to thank Bernaurdshaw Neppolian for travelling to the UK to participate in the July 2023 event and for sharing his insights and experience. The project team would like to thank The Higher Education Innovation Fund at the Open University for financial support of the three events described above. We would also like to thank the OU Research and Enterprise Team, the OU Science, Technology, Engineering and Mathematics Faculty and the leadership of the OU School of Engineering and Innovation for advice and assistance with the HEIF funded workshops and Professor Nuttall's hydrogen research at the OU more generally. While this book is informed by the HEIF funded workshops, its production has been a separate project.

Finally, noting that currently hydrogen energy is a very fast-moving field, innovations and policy initiatives linked to the key themes identified in the stages described above have been amplified based upon approximately one year's news feed monitoring. These sources are referenced in the text.

The authors would like to acknowledge Marc Cochrane for his help in preparing the section relating to the rainbow colours of hydrogen and wider assistance.

The authors are most grateful to The University of Houston for financial support of this publication. That support has made possible the open access distribution of this work. The authors are also most grateful to Springer-Nature, our commissioning editor Anthony Doyle, Amudha Vijayarangan for book production assistance and their various colleagues for help and support.

This book has been prepared by William J. Nuttall, Karim L. Anaya-Stucchi, Adetokunboh.T. Bakenne, Joe Powell and Andrew Wilson. The authors alone are solely responsible for its content and the varying opinions expressed do not necessarily reflect those of The Open University, The University of Houston, the publisher or any contributing organisations or individuals except where explicitly acknowledged. Some third-party contributions have kindly been supplied for Chap. 5. Any errors or omissions concerning this book are the responsibility of the authors alone.

May 2024

Open Access Sponsor

The University of Houston (UH) Energy Transition Institute (ETI) has contributed staff time to the production of this book and generous funding so that the resulting eBook can be read at no cost by all. The authors gratefully acknowledge the generous support of UH ETI in making Open Access possible. The UH ETI advises:

At UH ETI, we are pleased to contribute to the variety of opinions and perspectives in the The New Hydrogen Economy with insights from the U.S. and Houston, a global leader in energy. Electrification will play a central role in global decarbonization, but hydrogen and its derivatives can offer transport and storage of clean energy to allow global access to regions of high resource intensity with lower costs and greater resilience, while also serving applications that require high energy and power density. As a molecular fuel, it can serve multiple sectors and functions that would otherwise be difficult to decarbonize. Hydrogen must earn its place in the future energy economy, regardless of production pathway, via safe and cost effective deployment of technologies and supply chains with zero or very low greenhouse gas emissions or fossil carbon intensity, with acceptable societal footprints. The evolving energy industry is committed to examining hydrogen's potential relative to competing options for decarbonization, and meeting global energy needs.

Notes

This book draws upon an earlier series of knowledge exchange meetings funded by the UK Higher Education Innovation Fund (HEIF) via The Open University, UK. That earlier project was focussed on information sharing between academia and industry. The HEIF funding was awarded for two project phases. The initial scoping phase (2019–2020) comprised an information sharing exercise between The Open University and Royal Dutch Shell. The relationship involved no transfer of funds.

While drawing upon lessons learned from the HEIF funded project, this book is a separate and distinct product. The journey to the book and its publication have been discussed in the acknowledgements.

This book is not a history of hydrogen as a chemical element and a technical proposition. Readers seeking such a reference are recommended to read:

Michaela Kendall, Kevin Kendall and Andrew P.B. Lound, (2021), *Hystory—the Story of Hydrogen,* Adelan ISBN 978-0-9561111-4-2

This work is intended to provide information and insight to those with some existing knowledge of hydrogen energy systems. While we have tried to write in an accessible style, it is not an introductory primer. For a more introductory treatment we recommend a previous book published by two of the authors of this work. That previous work is:

William J. Nuttall and Adetokunboh T. Bakenne, (2019), *Fossil Fuel Hydrogen-Technical, Economic and Environmental Potential,* Springer ISBN 978-3-030-30907-7

Disclaimers

Liability: The authors, other contributors and the publisher cannot assume responsibility for the validity of all materials or for the consequences of their use. This book contains no advice or guidance and should not be used as the basis of any investment or other decision.

Trademarks: Product or corporate names may be trademarks or registered trademarks, and are used only for identification and explanation without intent to infringe.

Images: Efforts have been made to establish and contact copyright holders for all images presented in this book. We are most grateful to all the various rights holders who have kindly granted permission for reproduction. Despite our best endeavours, there may be instances where the rights of third parties have been overlooked. In such cases, we apologise and we ask that rights holders make contact and we will endeavour to resolve matters.

Contents

About the Authors

William J. Nuttall has more than 20 years' experience as a researcher and commentator on energy technology and policy. Originally trained in physics (MIT USA PhD 1993) his main areas of professional interest have been nuclear energy in all its various aspects, and hydrogen energy. He is the author with Dr Adetokunboh Bakenne of *Fossil Fuel Hydrogen* (Springer, 2020). From 2012 to 2024 Dr Nuttall was Professor of Energy at The Open University, UK. He is currently Professor of Energy Policy and Transformation in the School of Electrical, Electronic and Mechanical Engineering (part time) at the University of Bristol, UK. He maintains close links to various universities and research institutes worldwide.

Joseph B. Powell is Executive Director & Shell Endowed Chair for Energy Transition at the University of Houston, and Professor of Chemical and Biomolecular Engineering. He is a member of the U.S. National Academy of Engineering, Fellow and former Director of the American Institute of Chemical Engineers, former Shell Chief Scientist—Chemical Engineering, former chair of the U.S. Department of Energy's Hydrogen and Fuel Cell Technical Advisory Committee, and served more than 36 years in industry in commercialization of new processes for production of chemicals, fuels, and biofuels, and global strategy for the energy transition to a net-zero carbon economy. He is co-inventor on more than 125 patent applications, has received AIChE/ACS/R&D Magazine awards for Innovation, Service, and Practice, and is co-author of *Sustainable Development in the Process Industries: Cases and Impact* (Wiley, 2010).

Dr. Powell obtained a Ph.D. in chemical engineering from the University of Wisconsin-Madison, following a B.S. from the University of Virginia.

Karim L. Anaya-Stucchi has over 15 years of experience in utility regulation, having worked for telecom, water and energy regulators in developing and developed economies. She is a former Senior Research Associate at the Energy Policy Research Group (EPRG) at the University of Cambridge. She has helped to shape the next generation of electricity networks and system operators (assessing different innovation projects and local flexibility markets), collaborating with regulatory agencies, public utilities and electricity system operators. She has also advised on price control regulation, providing new and enhanced evidence of productivity growth of GB electricity distribution networks. She is currently a Senior Regulation Lead at ENOWA, an energy and water utility. Here, she is helping to build the new energy pricing and carbon regulation frameworks, electricity tariff scheme and energy market design for NEOM, a special economic region in Saudi Arabia. Karim holds a PhD degree in Energy Economics and a Master degree in Technology Policy from University of Cambridge.

Adetokunboh T. Bakenne is an experienced engineering researcher and consultant. He obtained his PhD in Nuclear Engineering from the University of Manchester. He then focussed on matters of energy technology and policy with his move to The Open University, where he worked as a research fellow on the future of the hydrogen economy. Bakenne has extensive consultancy experience in matters concerning commercial technology and risk assessment in a range of sectors.

 **Andy Wilson** is an experienced engineer with extensive experience in the energy sector. He obtained his PhD in Nuclear Engineering from The Open University, in the areas of light water reactors, hydrogen production and energy storage . He has since worked for the Civil Service in energy storage and is currently working in the nuclear fusion sector.

Glossary of Acronyms

AC	Alternating Current
ACT	Advanced Clean Trucks
ACTL	Alberta Carbon Trunk Line
AGHA	Africa Green Hydrogen Alliance
AGR	Advanced Gas-cooled Reactor
ARENA	Australian Renewable Energy National Agency
ATR	Autothermal Reforming
BAU	Business as Usual
BCM	Billion Cubic Metres
BECCS	Bio-Energy with Carbon Capture and Storage
BEV	Battery Electric Vehicle
BOEM	Bureau of Ocean Energy Management
BP	British Petroleum
CAGR	Compound Annual Growth Rate
CAPEX	Capital Expenditures
CCC	Climate Change Committee (UK)
CCGT	Combined Cycle Gas Turbine
CCS	Carbon Capture and Storage
CCU	Carbon Capture and Utilization
CCUS	Carbon Capture, Utilization and Storage
CI	Carbon Intensity
CO	Carbon Monoxide
CO2	Carbon Dioxide
COAG	Council of Australian Governments
CU	Carbon Utilization
DAC	Direct Air Capture
DC	Direct Current
DEWA	Dubai Electricity and Water Authority
DOE	Department of Energy (USA)
EOR	Enhanced Oil Recovery
ESG	Environmental, Social and Governance

EU-ETS	European Union—Emissions Trading Scheme
EV	Electric Vehicle
FC	Fuel Cell
FCEV	Fuel Cell Electric Vehicle
GCC	Gulf Cooperation Council
GHC	Green Hydrogen Coalition
GHG	Greenhouse Gas
GT	Gas Turbine
GTL	Gas to Liquid
GWP	Global Warming Potential
H2	Hydrogen
H2H	Hydrogen to Humber
HDT	Heavy Duty Truck
HEIF	Higher Education Innovation Fund
HEV	Hybrid Electric Vehicle
HHV	Higher Heating Value (includes heat of vaporization for the water in the fuel)
HICE	Hydrogen Internal Combustion Engine
HTGR	High Temperature Gas-Cooled Reactor
HTS	High Temperature Superconductivity
HYCO	Hydrogen-Carbon Monoxide
ICE	Internal Combustion Engine
IEA	International Energy Agency
IOC	International Oil Company
IRA	Inflation Reduction Act (USA)
IRENA	International Renewable Energy Agency
IT	Information Technology
LCFS	Low Carbon Fuel Standard
LDES	Long Duration Energy Storage
LH2	Liquid Hydrogen
LHV	Lower Heating Value (excludes heat of vaporization for the water in the fuel)
LMP	Lithium Metal Polymer
LNG	Liquefied Natural Gas
MEP	Midlands Engine Partnership (UK)
MRI	Magnetic Resonance Imaging
NETL	National Energy Technology Laboratory
NG	Natural Gas
NO	Nitrous Oxide
NOC	National Oil Company
NORM	Naturally Occurring Radioactive Materials
O&M	Operations and Maintenance
OCAP	Organic CO_2 for Assimilation by Plants
OECD	Organisation for Economic Co-operation and Development
OEM	Original Equipment Manufacturer

OIES	Oxford Institute for Energy Studies
OPEC	Organization of Petroleum Exporting Countries
PEM	Proton Exchange Membrane
PHEV	Plug-In Hybrid Electric Vehicle
PE	Piston Engine
PGM	Platinum Group Metals
POX	Partial Oxidation
PV	Photovoltaic
SAF	Sustainable Aviation Fuel
SMR	Steam Methane Reforming
SOEC	Solid Oxide Electrolysis Cell
SOFC	Solid Oxide Fuel Cell
STP	Standard Temperature and Pressure
SUV	Sports Utility Vehicle
TCO	Total Cost of Ownership
WGS	Water Gas Shift
WTI	West Texas Intermediate

List of Figures

List of Tables

Chapter 1
Introduction and Background

Abstract The origins of the book are introduced. The book derives from a knowledge exchange set of meetings primarily between industrial experts and academics. The chapter introduces the principal issues around hydrogen as an energy carrier. In particular its potential role in a low carbon energy system is described. The concept of the 'rainbow colours of hydrogen' is introduced and key colours of hydrogen denoting various production methods are explained.

Hydrogen has long been lauded as an important energy carrier and key enabler in the world's shift to a low-carbon future. Despite this, and despite the fact that hydrogen production technology has long been available, the much sought after 'hydrogen economy' has been elusive, and subject to a number of false starts. Recent years have seen renewed interest in low-carbon hydrogen as a path for difficult-to-decarbonise sectors to follow.

Broadly, it is the starting assumption of this book that there are two likely candidates for hydrogen production; Green Hydrogen (hydrogen produced from renewable electricity) and Blue Hydrogen (hydrogen produced from fossil fuels, primarily natural gas, with carbon capture and storage). Whilst these would appear to be two very separate paradigms, the reality is far more complex. The different characteristics of the two approaches present a range of potential benefits to wider industry and consumer use of hydrogen (and the by-products of its production).

In 2020 and 2021, the COVID-19 pandemic resulted in significant disruption to the international oil and gas industry. The reduction in transport usage globally and the resulting collapse in the oil price had significant implications for the sector. Since the pandemic, many businesses have placed a greater emphasis on remote working and limiting travel, especially international travel (Office of Tax Simplification 2022). Looking further ahead one can expect, for key territories at least, that post-pandemic plans will align with a lower carbon agenda motivated by concern to mitigate global climate change. While such shifts might be regarded as problematic for the oil and gas industry, this situation represents a rare opportunity for the broader energy industry to reconsider its market strategy at a fundamental level. Such a shift in thinking might

W. J. Nuttall et al., *Insights into the New Hydrogen Economy*,
https://doi.org/10.1007/978-3-031-71833-5_1

yield a much more positive attitude to clean hydrogen energy. There are indications that such a shift is already underway.

This book is the output of a process of knowledge exchange primarily between industry and academia. The industry-academia interactions occurred in the period 2020–2022. The first phases of the activity were greatly affected by the COVID-19 global pandemic. By necessity most discussions and the initial events were conducted remotely. The benefit, however, was that we are able to develop a global overview of the issues with partners joining us from North America and more widely.

The book is structured as follows: Chapter 1 introduces the origins of the book and provides an introductory commentary on hydrogen today. Chapter 2 describes the hydrogen industry of today and the drivers of change. Chapter 3 present major near-term opportunities for the expansion of the hydrogen industry. Chapter 4 examines progress in specific regions of the world. Chapter 5 presents a set of innovations and ideas that could be of great significance for the future trajectory of the hydrogen industry. The conclusions of the book are presented in Chapter 6.

1.1 Overarching Objectives

At the start of the 2020s there was a resurgence of interest in the potential for hydrogen as a low carbon energy carrier. There was a growing awareness that hydrogen could represent an opportunity to displace fossil fuel combustion. Such thinking was a dominant driver of academic interest. Meanwhile the natural gas industry was increasingly turning to the idea that natural gas to hydrogen conversion, coupled with carbon capture utilisation and storage (CCUS), could be a way to preserve, and indeed even strengthen, their industry consistent with a low carbon future (Global 2023). At the start of the project, the dominant questions being addressed by UK academia and public policy were focussed on the role of hydrogen in a renewables-based energy system with very high levels of electrification. Over the last three years the balance of concern has shifted somewhat, but in this book, we suggest that the way ahead for hydrogen utilisation at scale can still only be understood through engagement with industrial expertise. This book seeks to advance such understanding.

A key guiding question at the start of the project was:

"Where in the world will material hydrogen production arise first, and why?"

In seeking to address this question, the study explored hydrogen developments around the world and the technology and policy issues driving change in different territories. Our opening question raised a set of complex and multifaceted issues discussed from different perspectives.

These different perspectives included: international energy companies, industrial gases companies, commodity traders, shipping companies, infrastructure companies and technologists.

The project noted the very significant levels of investment needed to enable a shift to hydrogen at scale. The project was interested in sources of venture finance and also for innovation strategies that included the disruptive and the incremental. The project also included the potential for science and technology led innovation, such as in hydrogen cryomagnetics and geological hydrogen both of which are discussed later in this book.

1.2 Origins of the Ideas Presented

This book follows on from a knowledge exchange activity led by The Open University and funded by the UK Higher Education Innovation Fund (HEIF). Despite its academic roots, this work is not an example of the following:

- **Traditional academic research**. The HEIF funded project was an information sharing exercise between industry and academia and it was not intended to generate new knowledge. It was instead intended to collate and assess existing knowledge and its implications on the hydrogen production and supply landscape. The HEIF project was intended to provide insights and ideas for policy-makers and industry strategists. It also provided pointers for future research within academia and elsewhere.
- **Direct strategy recommendations for any one stakeholder**. The focus was on the strategic options faced by the energy sector as a whole and the role of policymaking in facilitating the emergence of hydrogen as a major commercial energy commodity. The scope and agenda for the HEIF funded activity were not geared to specific commercial interests of any energy companies that provided advice or assistance at that time.

Although this book was made possible by the earlier HEIF funded project (2021–2022), it is formally separate from it.

1.3 Knowledge Transfer Meetings Held in 2021 and 2022

This project has at its heart a series of expert dialogue events conducted in a spirit of knowledge exchange primarily between industry and academia. This project has sought to amplify the communication of ideas and insights from industry into academia. This book is based on two virtual expert gatherings conducted during the COVID-19 global pandemic in 2021 and an in-person event in London held in July 2022.

The first project webinar was held on 19 March 2021 under the title *Ports and Pipelines – Key to Britain's Hydrogen Future.* The four panellists were all industry experts with experience in: major international oil and gas companies, a leading global commodities company, a prominent industrial gases company. The focus of the event was on hydrogen infrastructures and the relationship to industrial clusters. The primary focus was on the United Kingdom.

The second webinar was held on 24 June 2021 under the title *Molecules Are Key to the Low Carbon Future*. In this case the geographical focus was on North America. This second webinar was kindly co-organised with the Payne Institute for Public Policy at the Colorado School of Mines, USA. A major theme was the competition between, and the synergies amongst, future expansion of hydrogen as an energy carrier and electrification of the energy system. The role of hydrogen in US energy innovation was specially considered. The event conveyed the message that hydrogen could have a very large role to play and that, in the USA, transport applications could be especially important. This second webinar also highlighted specific aspects of hydrogen innovation that are usually not addressed in more introductory approaches. Such issues included – private sector innovation leadership, geological hydrogen, and hydrogen and superconductivity.

All three events had an expert audience. The first event involved approximately 35 participants and the second Involved approximately 50 participants with the wider participation reflecting the broadened transatlantic focus. The third event was held in person in London in 2022 and it involved approximately 30 participants.

1.4   Hydrogen in Context

Most of the nations of the world have accepted the need for action on climate change. In recent years, many nations have seen significant investment in renewable sources of electricity. As a result, much of the 'easy' decarbonisation of energy consumption is underway. Means of producing clean electricity are well understood and have already seen significant investment and policy support. What is more, the majority of the existing infrastructure is well suited to at least partial penetration of renewable generators. Aside from the not insignificant need for energy storage system investment, it would appear that the electricity consumption has a relatively clear path towards decarbonisation.

This is, of course, only part of the story: electricity today comprises a little over a third of energy consumption in the USA, as Fig. 1.1 (taken from Lawrence Livermore National Laboratory (2019)) shows. The remaining two-thirds of energy consumption is in transportation and heating, with the majority being direct consumption of fossil fuels. If this portion of energy use is to be decarbonised, it will be either via electrification or via the use of some sort of molecular energy carrier. The reality is likely that there will ultimately be a combination of both. For the latter, hydrogen has for decades been seen as a key candidate as a low-carbon molecular energy carrier. There is, however, still no broad consensus on how it should be produced at scale, how it might be supplied, and which customers might be the earliest adopters.

Hydrogen has long been an important chemical to the fossil fuel and ammonification industries, which are responsible for nearly 90% of its consumption today. The majority of hydrogen today is produced from fossil fuel sources and is, in many cases, a carbon-intensive process. The leading method for production is SMR (steam-methane reformation), where methane in catalytically reacted with steam to produce

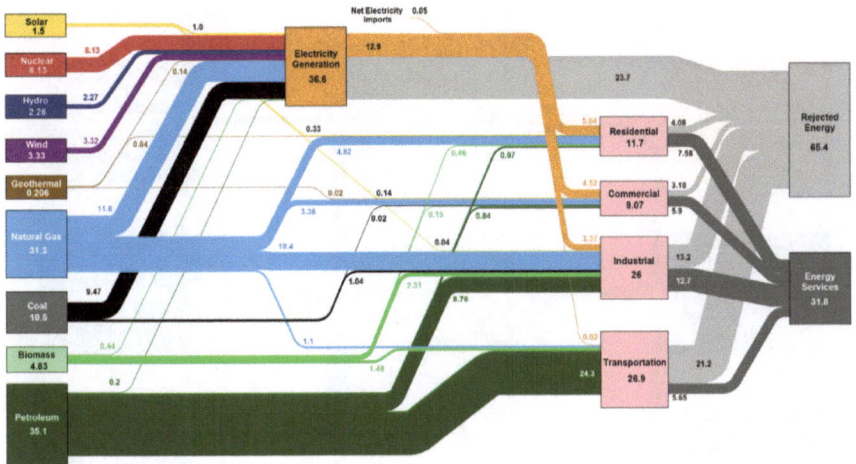

Fig. 1.1 US Energy Consumption in 2021 as estimated by Lawrence Livermore National Laboratory. For assumptions see: https://flowcharts.llnl.gov/. Source: US-DOE

hydrogen. Not only do the reactions themselves produce carbon dioxide, but so too does the burning of methane - as required to supply heat to an endothermic process.

As Professor Neppolian Bernaurdshaw advised the authors: hydrogen has the highest energy density (~120 MJ/kg) of any candidate fuel for mainstream use (Møller et al. 2017). In addition, Dr Michaela Kendall advises the project that Hydrogen fuel can be 65% cheaper than diesel and can bring lower carbon footprint too (85% reduction). She has also cautioned that a key factor in the slow roll out of hydrogen thus far has been the challenge of establishing supply chains. She has further observed that the main issue thus far has been government policy, but that a transition to market led dynamics appears now to be in prospect.

1.5 Decarbonising Hydrogen Production

There are, in a broad sense, two approaches to decarbonised hydrogen production. Green Hydrogen is produced from a renewable source. Since the majority of renewable sources generate electricity, this is generally done by some form of water splitting by electrolysis, either room temperature water electrolysis or higher temperature steam electrolysis. For a comprehensive discussion of current electrolyser technologies and those in development for Green Hydrogen, see Patonia and Poudineh (Patonia and Poudineh 2022). It is worth noting here that hydrogen is only one product of this process; oxygen is also produced, but it is typically vented to the atmosphere. Blue Hydrogen is produced from fossil fuel sources, usually natural gas, and the process is integrated with some form of CCS (carbon capture and storage) or CCU (carbon capture and utilisation). Whilst the aforementioned SMR process is the

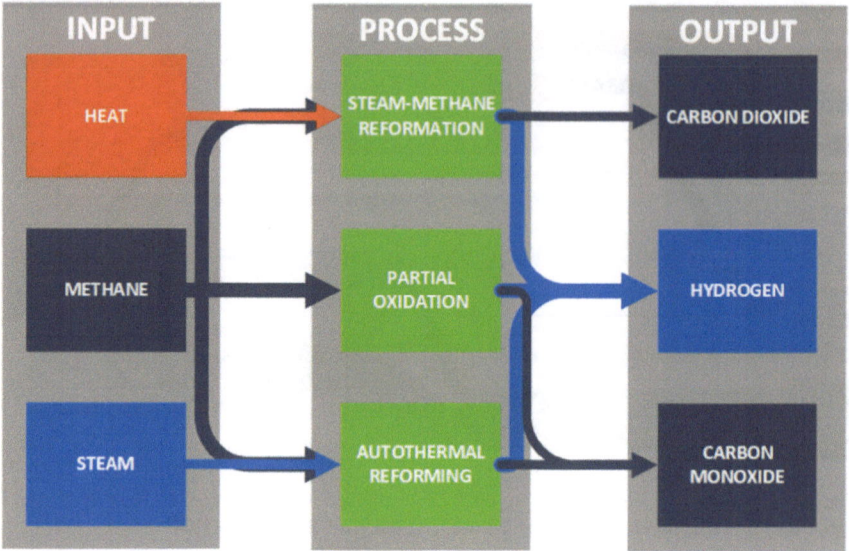

Fig. 1.2 Routes for fossil-fuel hydrogen production. Authors' own work

prevailing process as of today, there are several alternative processes including POX (partial oxidation) and ATR (autothermal reforming). Each has a different set of characteristics, requires different inputs, and yields different products, as summarised in Fig. 1.2.

CCS involves separating the carbon products from the waste stream associated with the consumption of a fossil fuel. Historically, this has meant separating the CO_2 from the flue gas exhaust of a fossil fuelled power plant. The energy penalty associated with separating the CO_2 from the other constituents of flue gas (mostly nitrogen) directly impacts the energy yield of the plant in question. In this light, then, the case for post-combustion power plant CCS would appear weak since it would increase costs and reduce revenue.

Blue Hydrogen provides an opportunity to reframe the hydrogen research agenda within academia. In the Europe Union and the United Kingdom the bulk of academic interest in hydrogen has been devoted to the issues of Green Hydrogen and to the relationship to the electrification of the energy system based upon new renewable sources of power generation.

The exhaust carbon streams from the hydrogen production processes shown in Fig. 1.2 are richer in carbon dioxide than flue gas streams from power plants, where CO_2 is very diluted. The relatively pure nature of these waste streams makes the carbon capture process significantly less energy intensive. There is an important caveat to this in the context of the SMR process, which requires the supply of heat via combustion of natural gas, whose exhaust stream has the lower CO_2 concentration of a flue gas. As a result, the ATR and POX processes, which offer an advantage over SMR in that they do not require this difficult to decarbonise supply of heat, provide additional opportunity and should be worthy of consideration. Blue Hydrogen's reliance on external sources of steam, oxygen or CO presents hurdles, but also synergistic opportunities with other industrial processes, and potentially with nuclear sources of heat.

Many remain sceptical of the merits of carbon capture and whether it provides a beneficial contribution for the world's present difficulties. Critics observe that this technology is in its early stages, it is energy intensive, and some would further argue that it is incompatible with a route to full decarbonisation. Others, however, argue that coupled with natural gas innovation it has the potential to reduce emissions rapidly and at large scale and if CCS is coupled with biomass innovation has the potential (through negative emissions) to greatly boost the planet's journey to Net Zero (Nuttall and MacGregor 2024). The fact that SMR yields a relatively pure carbon dioxide waste stream, means that it lends itself to the use of CCS and reduces the energy lost per unit of carbon captured when compared to the capture of carbon from, for example, the flue gas from a natural gas power plant. CCS is only one side of the carbon capture coin. Greater all-round benefits could be garnered via the utilisation of captured carbon by industry, in the form of CCU. Carbon monoxide is in itself an important industrial reactant. In re-contextualising carbon as not an expensive waste stream liability, but as a potential revenue-generating product, better decisions might be made as to hydrogen production process suitability. In principle, Blue Hydrogen production can be adjusted (via the addition of externally supplied oxygen or heat, for example) to yield essentially pure CO_2 waste stream further increasing the attractiveness of CCUS.

In the early 2020s, there has been increased interest in CCS (Evans-Pritchard 2020). Arguably the US led the world in deploying its 45Q tax credit that rewards the use of CCS in major projects, see Chapter 4 for further details. The then UK Prime Minister's previous opposition to the technology turned into grudging support:

"We want to lead on carbon capture and storage, a technology I barely believed was possible, but I am now a complete evangelist for" – Boris Johnson (YouTube 2020).

This is also aligned with his announcement made in April 2022 to double the UK's expected low carbon hydrogen capacity by 2030 (E.a.I.S. British energy security strategy 2022).

The major oil and gas companies, in arguably facing an existential threat in the form of a looming climate crisis have also, with varying levels of enthusiasm, come to support this technology, once thought to be too expensive, as a means to maintain their role in the energy industry:

"The oil and gas fraternity has embraced carbon capture with the zeal of the converted, betting that this neglected technology can be made cheap enough, quickly enough, to head off the seemingly unstoppable march towards electrification and the Green Hydrogen economy." Ambrose Evans-Pritchard (Daily Telegraph) (Evans-Pritchard 2020).

The IEA has been similarly supportive of CCS seeing it, and Blue Hydrogen, as key technologies supporting a transition to a hydrogen economy (International Energy Agency 2019).

In early March 2024 Shell and ExxonMobil jointly announced plans to work together and with the government of Singapore to create a CCS hub in the city-state (ExxonMobil 2024). The project, known as S-Hub, has the potential by 2030 to capture and permanently store at least 2.5 million tons of CO_2 annually.

Some might argue that the international oil companies are indeed transforming their business model towards a low carbon future, but the direction of their strategy is electrification, not hydrogen. As authors of this book, we observe, however, that the future remains far from clear but what does appear certain is that the oil industry, or what might become the former oil industry, remains very interested in hydrogen as a strategic option and indeed there is reason to believe that the arguments in favour of hydrogen being a major part of the future for today's oil majors is looking more likely every day. Key to this thinking is the energy transport and storage advantages of hydrogen coupled with the competencies of the incumbents – arguably an international oil company knows more about hydrogen than electricity. These advantages can potentially outweigh the clear efficiencies (for example in energy conversion) offered by the electricity alternative. As the energy transition shifts from being a technical question to a question in business strategy and political economy we posit that the relative importance of hydrogen over electricity will grow. That said, we must be clear. The future will see a massive growth globally in – electrification, renewable electricity generation, and electricity storage. But we further posit that there is an even stronger growth potential for an evolutionary expansion of today's hydrogen industry into multi-trillion-dollar global energy system with low carbon production, shipping, trade and end use. In this context, we see a bright future for both Green Hydrogen and Blue Hydrogen.

For the international oil companies the push for hydrogen is not an attempt to hang on to something that is not optimal or useful as some claim, but rather we suggest it is a competency that can be leveraged relatively rapidly and in an evolutionary way into services that are most useful and substantially more sustainable than today's fossil fuel combustion.

In the first webinar held in March 2021 it was emphasised that hydrogen production opportunities span far more than the much-discussed alternatives of Blue and Green. Rather than focussing on the rainbow colours of hydrogen (see Sect. 1.10) we tend to the view that if a global manufacturing and trading regime can be established in which hydrogen fully covers economically its environmental damage at the point of manufacture then, in distribution and retail sales policy, markets should not distinguish between hydrogen offerings based upon upstream manufacturing procedures. To develop the hydrogen economy based upon the hydrogen carrying with it an explanation of its origins would require end-to-end certification and proofs of provenance and origin. This would be for a product where essentially it is impossible to distinguish with certainty the origin of the product based upon its chemical composition (impurities). It is quite possible to change the impurities in the product with post-production processing. From this project the suggestion emerges that hydrogen production should face charges (such as a carbon tax or permit price) appropriate to the environmental harm involved in the production process. Once such measures have been taken at the point of production no further distinction should be made – hydrogen is simply hydrogen. We shall return to this point in Chapter 6. Care will however be needed concerning international trade in the cases where receiving territories, for whatever reason, find themselves unable to trust upstream arrangements. In that case border tax adjustments, or even stronger measures (such as restraints on trade) may be needed.

The second webinar learned of the successful use of public policy power in California and the United States more widely. As will be discussed further in Chapter 4 (section 4.1.1), the US Federal 45Q tax credit has already had much impact and as a consequence of its incentivisation of CCS, steam methane reforming of natural gas for hydrogen production with CCS (i.e. Blue Hydrogen) is already a potentially profitable activity without further public policy intervention. More recently the Biden Administration's Inflation Reduction Act signed into law in August 2022 provides a significant boost to hydrogen industrial development. Two key measures are included. The first is an adjustment to the US tax code (known as 45 V) which will be based upon the life cycle emissions of clean hydrogen production and hence will incentivise progress towards Net-Zero, this is accompanied by a strengthening of the existing 45Q tax credit for CCS, discussed later in Chapter 4. The other measure is an allocation of $8 billion for a series of hydrogen hubs. That measure is expected to support the roll out of infrastructures associated with hydrogen production using renewable energy, natural gas reforming with CCS and hydrogen produced using nuclear energy.

Figure 1.3 (taken from California Air Resources Board (2021)) shows a waterfall plot for cost of CO_2 capture from industrial facilities. Capture of CO_2 from bio-ethanol fermentation off-gas, or from hydrogen production via SMR when combined with the California Low Carbon Fuel Standard (LCFS) and the original U.S. 45Q tax credits for CO_2 storage can provide a positive rate of return, and hence are attractive current targets for industry.

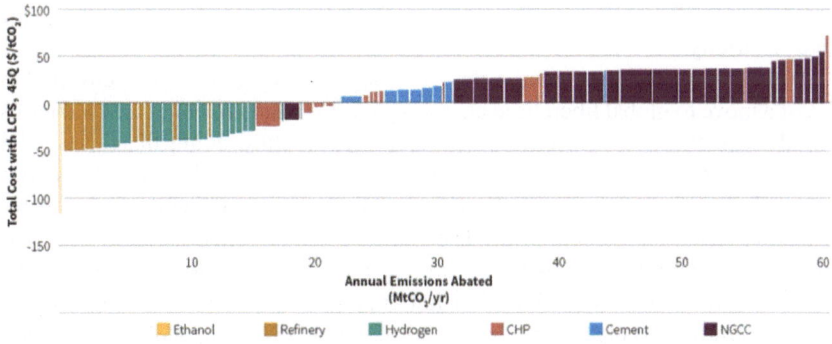

Fig. 1.3 Marginal abatement curve by facility. California Air Resources Board 2021 Source: Energy Futures Initiative and Stanford University, with kind permission

In this book we do not consider hydrogen through the lens of "sustainability" however that may be defined. Indeed, it is arguably the ambiguity of the various definitions of such ideas that leads us to avoid consideration of "sustainability" as a motivating goal. Rather, we restrict our sense of technological merit to being defined via a particular set of core considerations:

- Engineering practicality, especially at scale and at speed.
- Cost and value.
- Direct whole life-cycle impacts, especially environmental impacts, of decisions made.

That is, in this book, we are of course focussed on what **should happen** – we are utterly focussed on the ambition of a Net Zero global future. We are also focussed on the technological limits of the possible, and what **could happen**. Finally, we recognise that solutions need to be affordable for society as a whole and we recognise the importance of economics in determining what **will happen** in the global energy market economy.

We acknowledge that the future of the global energy market looks very different today than it did 50 years ago. In the late twentieth century oil was a genuine energy commodity (Madhavi and Nuttall 2023). Crude oil was energy dense and easily transported and, as such, landed costs differed little in different parts of the world. The world could operate with, in effect, a single global oil price. Key benchmarks indicative of that price included West Texas Intermediate (WTI) originating in the USA or, in Europe, Brent Crude. For decades US foreign policy and military power protected the global oil market. The US Navy patrolled regions of the world key to oil supply (such as the Persian Gulf) even though very little such oil made it to American shores. In essence American hard power was protecting the price and the market rather than specific cargoes. It was widely believed in energy policy circles that natural gas would join oil as a genuinely traded market commodity led by the fact that increasingly natural gas would be supplied as a maritime distributed liquid product (liquefied natural gas) rather than in the rather more traditional (and geographically restricted) pipeline commodity that it was initially. Markets were even developed for something that remains a largely non-storable commodity – grid scale electricity.

Over the last twenty, or more, years, developments have, however, subverted the idea of a move to global liberalised all-energy market. The growth of renewables has shifted consideration away from fuels towards infrastructure, noting that there are few things more political and less economic than infrastructure. Perhaps more important the very foundation of energy markets is crumbling – a global liberal and transparent oil market based on a single price. The war in Ukraine which flared in 2022 with renewed Russian invasion has led to large opaque oil trades, for example between Russia and India (Reuters 2023a). In addition, a key area of global oil demand, the European Union, has placed price caps on Russian oil below the global market price (Politico 2023). US criticism of such moves has been remarkably muted. Hence, one might infer that the highest US priority is no longer the preservation of a global oil market, but rather the ending of Russian aggression is seen as a higher goal.

More broadly perhaps the US is now looking ahead to an entirely new geopolitics of energy. Arguably such a shift would be as profound a part of the energy transition as decarbonisation. In that scenario hydrogen could be a key enabler of such a shift. The visceral national desire in US politics to end the 'forever wars', many of which have been associated in some way or another with oil, is of key importance when examining US attitudes to the energy transition.

Despite the weakening of global energy markets and a possible drift towards a world of secret bilateral contracts, the global oil market is not yet dead. For the rest of the book, we shall imagine a role for hydrogen as seen in global market terms. As with oil and LNG, hydrogen could be shipped far and wide and traded globally. Some parts of the world could be producer regions while others will be consumers. We suggest that such a global market will be at scale and be a positive step towards a low carbon future. We posit such a future ahead of notions of local self-sufficiency, but we do not discount the importance of such ideas and actions. We see self-sufficiency as being a key concept, but not the only concept, underpinning notions of energy security. The authors of this book see energy security as arising from both energy independence and resilient global markets. It is the idea of a resilient global hydrogen market that we shall use as a backdrop in what follows.

1.6 Customers

A thriving hydrogen market will of course be dependent on customers. The potential applications for hydrogen are numerous. This subsection will note the highest profile and nearest-term applications for hydrogen.

Hydrogen could play a key role in transportation in the future, in particular in FCEVs (fuel cell electric vehicles). Many of the roadmaps cited in this document suggest that hydrogen is well suited for roles that batteries have so far struggled to fulfil, such as some routes and services for buses, coaches, and heavyweight haulage. Trains are also a potential end-user in cases where overhead or third rail electrification is problematic. In all of these cases, the vehicles are heavy and have high power needs. Hydrogen is a major enabling fuel with zero tailpipe emissions (when used in FCEV, noting the risk of NOx emissions in hydrogen ICE use). A key virtue is its energy density and the possibility of rapid refilling, making it preferable to batteries in the majority of cases, especially in those cases where journeys are relatively infrequent or long-distance rendering more difficult the provision of appropriate electricity infrastructure.

Concerning passenger vehicle road transport. Here, hydrogen faces greater competition from battery electric cars than in any of the heavy transport use cases discussed above. Despite this, there is already at least one hydrogen vehicle success story. In California, an abundance of renewable electricity has enabled a small but growing network of Green Hydrogen filling stations and some adoption of hydrogen FCEVs (California Energy Commission 2019). Future plans include a desire to stimulate the adoption of FCEVs for the taxi and ride-hailing industry.

Hydrogen is also suitable for blending into existing natural gas networks. In the UK, consensus is that at concentrations at, or below, 20% hydrogen by volume, there is no general need to alter either existing distribution networks or consumer's home boilers (Global 2022). In other territories the 20% figure is somewhat contested. In its 2020 the US DOE published its hydrogen strategy which also pointed to the possibility of 20% blends in some contexts (Department of Energy 2020). This presents an opportunity to reduce emissions in the difficult to decarbonise sector of home heating and is exactly the goal of the UK's HyNet project (HyNet 2018). A more ambitious option is the conversion of a natural gas network entirely to hydrogen. This requires significant changes to both the natural gas grid and consumer devices. The payoff is greater decarbonisation of those sectors of the gas grid. This, for example, was the goal of H21's Leeds City Gate project (H21 2019) which has since grown into wider UK hydrogen ambitions, see Chapter 4.

1.7 Established Approaches to Production

The contrast between the two approaches (green and blue) is not simply in the approach to production; Green Hydrogen produces oxygen, whilst Blue Hydrogen can produce carbon monoxide. Both are potentially useful, and more importantly revenue-generating, given the right customer. Ultimately, it is not inconceivable that what is today considered a waste product could ultimately be the deciding factor as to which method of hydrogen production is most suitable for a given market or locale. Thus, there can be considered to be three key factors influencing the approach taken to hydrogen production:

- The **primary energy source** for hydrogen, be it renewable electricity sources or fossil fuels.
- The **infrastructure** to produce hydrogen and to supply hydrogen and additional products to end users.
- The **customer** for hydrogen and their potential needs for additional molecular products (such as O_2, CO or CO_2).

These factors are all likely to influence the **process** by which hydrogen is produced, be it blue or Green Hydrogen. The access to a primary energy source, the existing infrastructure (or otherwise) for supply hydrogen to a customer, and the customer themselves, are all contributing factors for assessing which process might be the most suitable.

1.8 Primary Energy Source

As discussed, the focus of this project is on two potential hydrogen sources: Blue and Green Hydrogen. This subsection will explore the nature of these two sources and speculate on the effect that their characteristics might have on their costs, their supply infrastructure, and their potential customers.

Green Hydrogen centres on the supply of abundant and cheap renewable electricity. The use of electrolysers has long been suggested as a potential means of load balancing intermittent supplies of renewable electricity. Green Hydrogen production and storage have a potential role in load-shifting electricity generation by storing excess energy in chemical form as hydrogen at times of reduced demand and releasing this energy at times of increased demand.

Initiatives to achieve large scale Green Hydrogen are proliferating worldwide. In Europe, HyDeal Ambition (hydeal 2022), an industry platform with the involvement of 30 energy players, intends to produce Green Hydrogen at scale with infrastructure developed across Western Europe: from Spain, Eastern France to Germany. Its target is to produce Green Hydrogen at €1.5/kg before 2030. HyDeal LA in California (Green Hydrogen Coalition 2022), which mirrors the HyDeal Ambition, aims to create a Green Hydrogen hub in Los Angeles Basin with a production cost of less than $2/kg by 2030. HyDeal LA is part of the Green Hydrogen Coalition (GHG) from USA that supports the deployment of Green Hydrogen for multisectoral decarbonisation across the whole supply chain.

Blue Hydrogen requires fossil fuels; natural gas is required for SMR and ATR, but POX is more flexible and able to rely on a range of fossil fuels. The ATR and POX processes also require additional feedstocks such as oxygen, although such issues are not a dominant concern for industry which tends to be more focussed on the capital investment in new units and capabilities than on the operational feedstocks they might require. Such a view is rational given that we are, in both cases, talking about the provision of oxygen as an additional input. At this point proponents of Green Hydrogen might validly point to the reality that the production of electrolytic hydrogen also generates an oxygen co-product that is usually considered a waste stream and vented. Hence one could imagine a hybrid solution combining elements of Blue and Green Hydrogen. The reality, however, is that for the established petrochemical industry the choice between, for example, an SMR and an ATR with the consequential need in the latter case for an oxygen supply the issue is quite simply not a major issue. It is frankly straightforward for those contemplating such an industrial facility to incorporate an air-separation unit for the supply of atmospheric oxygen if oxygen is required. That decision forms part of their capital expenditure thinking and is not really seen in terms of an operational feedstock cost. An ATR generally requires an external oxygen input. Through the additional of external oxygen relatively concentrated CO_2 emissions can be achieved. The resulting CO_2 has the potential to be very well matched to the needs of carbon capture and utilisation (CCU).

The source of hydrogen is a major factor contributing to its cost. The cost to produce hydrogen will be dependent on the price of its source: for Blue Hydrogen the fossil fuel price and for Green Hydrogen the electricity price. These considerations perhaps imply that the costs of Green Hydrogen production are likely to vary more, and be less predictable, than the costs of Blue Hydrogen – at least over shorter timescales.

Long duration energy storage (LDES), the cost of which will be reduce over time, will play an important role in Green Hydrogen production by dealing with electricity price volatility and by matching more efficiently electricity demand from electrolysers and supply from renewable sources such as wind and solar. There are some initiatives that are incorporating LDES as part of hybrid solutions, most recently for power purchase agreements (PPA). Innovative PPA arrangements can help to deal with uncertainties about matching power supply and demand (from renewables), and electricity price. A report from the Long Duration Energy Storage Council (Company 2022) suggests that a 24/7 "clean PPA" would help to match supply and demand for renewable more precisely by considering smaller time intervals (e.g. 1-hour granularity) when measuring electricity consumption and its carbon footprint. This would be possible by combining renewable power (with current PPAs achieving between 40–70% emission reduction of actual electricity consumption) with energy storage, avoiding the use of fossil fuel to fill gaps when the wind does not blow, or the sun doesn't shine. Higher costs of LDES technologies, and a lack of agreed standards, are currently among the main barriers for large-scale LDES adoption, but innovation is occurring rapidly.

1.9   Hydrogen From Renewable Sources

Green Hydrogen production has traditionally been dominated by electrolysis in its various forms including high-temperature steam electrolysis. Today's leading electrolyser technologies are either alkaline or proton exchange membrane (PEM) based. A full range of current electrolyser technologies is summarised in Table 1.1.

Professor Neppolian Bernaurdshaw from SRM-Institute of Science and Technology in Chennai, India kindly spoke at the third workshop held in London in 2022. (SRM-IST has its origins in Sri Ramaswamy Memorial Engineering College founded in 1985.) Professor Neppolian looked to the future of low-carbon hydrogen production associated with renewable energy. His suggestions concerning methods of interest is summarised in Table 1.2.

While the potential for new technologies is of great importance, it is not the purpose of this book to focus on the current state of the art in scientific innovation. Our focus is to explore the large-scale issues and trends shaping the hydrogen industry globally in the near-term.

Professor Neppolian also spoke about the potential importance of hydrogen in India. That is a fascinating topic meriting a study in and of itself. Professors Neppolian and Nuttall hope to address such issues in a joint publication in 2025.

Table 1.1 Current and near-term hydrogen production electrolyser technologies. Source: Mittelsteadt et al. (2015)

Type	Alkaline	Acid	Acid polymer electrolyte	Alkaline polymer electrolyte	Solid oxide
Charge carrier	OH^-	H^+	H^+	OH^-	O^{2-}
Reactant	Water	Water	Water	Water	Steam
Electrolyte	Na^- or KOH	H_2SO_4 or H_2PO_4	Polymer	Polymer	Ceramic
Electrodes	Nickel	Ir and Pt	Ir and Pt	Ni and Ag	Nickel cermet
Temperature	80 °C	150 °C	80 °C	60 °C	Approx. 500 °C

Table 1.2 Possible future renewable energy routes to low carbon hydrogen production. Source Prof B. Neppolian, with kind permission

Possible Future Renewable Energy Routes to Low Carbon Hydrogen Production	
Thermal	Thermolysis
	Thermo-catalysis
	Thermochemical
Electrical	Electrolysis
	Plasma are decomposition
Photonic	Photo-Electrochemical
	PV electrolysis
	Photocatalysis
Biochemical	Light/dark Fermentation
	Direct–Indirect Bio-Photolysis

1.10 Rainbow Hydrogen

This section, and Fig 1.4, draw upon unpublished work by Marc Cochrane. The authors are most grateful to him for his assistance but the authors alone are responsible for any errors or omissions.

The authors foresee a future in which hydrogen is supplied commercially and at scale. We propose that hydrogen should be a market commodity underpinning a low carbon future. Hydrogen entering the market should be generated without carbon dioxide emissions or, at a minimum if some level of greenhouse house gas (GHG) emission is unavoidable, then hydrogen producers must pay whatever charges society deems necessary to mitigate environmental harm. Such charges should not be less than the social cost of carbon (Martin et al. 2021) and should be framed with respect to NetZero 2050 policy goals. The charges should be sufficient to ensure a move to general societal decarbonisation which can be expected to rise with time, noting, for example, the prevailing lowest cost of atmospheric carbon removal via direct air capture (DAC), or equivalent 'backstop' approaches. These back-stop costs will become more important as global Net-Zero emissions is finally approached mid-century. Hydrogen production and distribution that is emissions free, or that more than covers its environmental impacts via supplementary levy appropriate to the times, should be welcomed into the market.

One can envisage that in the future hydrogen might be generated from renewable sources (e.g. solar or wind generated electricity or from renewable biomass pyrolysis), from fossil fuel sources with carbon capture and storage, from nuclear energy or be obtained directly in the form of geological hydrogen. All these low carbon sources, must more than off-set their life-cycle environmental impacts, but once that is achieved, the hydrogen should be admitted to the market without distinction. The hydrogen molecules will enter pipelines and storage systems without a need to record provenance nor data concerning its production history.

Many voices, commentating on the future of hydrogen, take a different view. For them the production and distribution of hydrogen has value implications beyond the value of the hydrogen itself and the direct environmental impacts. There is a sense that there is a hierarchy of virtue based upon engineering choices made rather than the product itself. All would agree that hydrogen is simply hydrogen however it is produced. There are those however who see hydrogen produced from renewable sources as being intrinsically more sustainable and hence intrinsically more valuable, and indeed virtuous. Such a hierarchy of virtue arguably gives rise to the widespread mention of "various colours of hydrogen".

Hydrogen is a colourless gas, is assigned a merely nominal colour depending on how the hydrogen is produced. It should be stressed that the allocation of colours to hydrogen can be a rather confused and sometimes inconsistent process. In what follows we do our best to respect the emerging orthodoxy, but references to other colour assignments can be found in the emerging literature.

1.10.1 Hydrogen from Hydrocarbons

Grey Hydrogen

Grey hydrogen is manufactured from natural gas or pure methane (Grid 2022; Newborough and Cooley 2020). The main processes used for Grey Hydrogen are steam methane reformation (SMR) and some autothermal reformation (ATR) both of which result in the production of hydrogen and carbon monoxide.

$$CH_4 + H_2O \rightarrow CO + 3H_2$$

This reaction is normally followed by a water gas shift reaction to convert the carbon monoxide to carbon dioxide, which also yields hydrogen.

$$CO + H_2O \rightarrow CO_2 + H_2$$

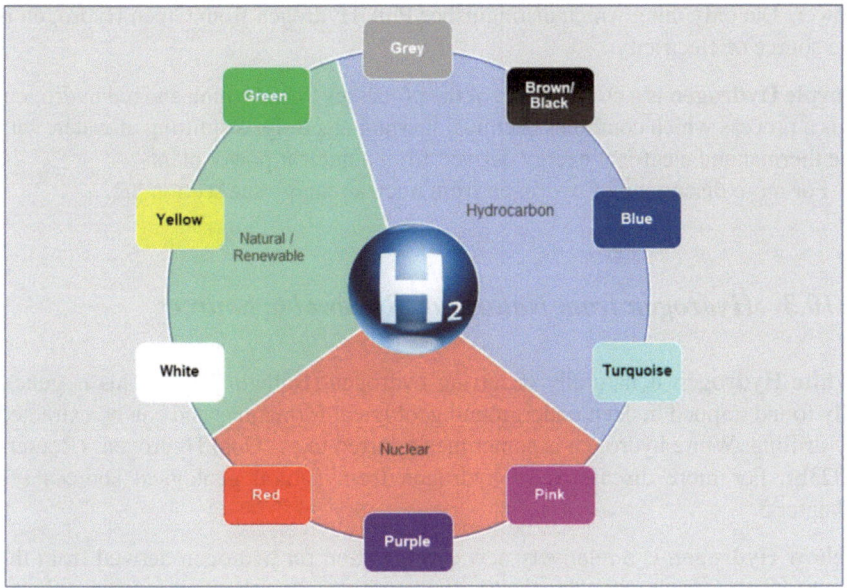

Fig. 1.4 The rainbow colours of hydrogen. Source: Marc Cochrane with kind permission

For grey hydrogen, the resulting CO_2 is vented to the atmosphere where it will contribute to anthropogenic climate damage. Just to note: Black and Brown Hydrogen are two terms for a process that starts with coal as a feedstock, as might be used to feed a HYCO-POX plant producing heat, hydrogen and carbon monoxide for industrial uses. As with Grey Hydrogen significant harmful greenhouse gas emissions can be expected.

Blue Hydrogen is like grey hydrogen, but in this case the resulting CO_2 is captured and stored geologically such that it should, ideally, not cause environmental damage – that is Blue Hydrogen relies upon Carbon Capture and Storage (CCS). Turquoise Hydrogen involves a high-technology process known as methane pyrolysis. It will be discussed in section 5.3.

1.10.2 Hydrogen from Nuclear Power

Red Hydrogen is produced from the high temperature catalytic splitting of water to produce hydrogen and oxygen. The high temperatures are achieved using nuclear power as a thermal energy source.

Pink Hydrogen is produced from the electrolysis of water, however the electricity used to split the water molecule into hydrogen and oxygen is derived from nuclear power. The only thing which distinguishes Pink Hydrogen from Green Hydrogen is the source of electricity.

Purple Hydrogen is a combination of the processes to make pink and red hydrogen. It is a process which combines chemical thermal electrolysis splitting of water, with the thermal and electrical energy derived from a nuclear power plant.

For more discussion of hydrogen from nuclear energy see section 5.2.

1.10.3 Hydrogen from Natural or Renewable Sources

White Hydrogen is naturally occurring hydrogen (Bulletin 2022). This is generally found trapped in deep underground geological formations and can be extracted by drilling. White hydrogen is sometimes referred to as 'Gold Hydrogen' (Reuters 2023b). For more discussion of hydrogen from natural geological sources see Chapter 5.

Yellow Hydrogen is a relatively new classification for hydrogen derived from the electrolysis of water when the electricity is produced by solar power.

Green Hydrogen is the term used to describe hydrogen produced with no greenhouse gas emissions. This is usually done by electrolysis of water using 'clean' electricity from surplus renewable electricity from one source or a combination of sources, such as wind, solar or wave power. The key feature is that no carbon dioxide is emitted from the electrical power generation (Grid 2022; Newborough and Cooley 2020; Bulletin 2022; Velazquez Abad and Dodds 2020). Electrolysis of water is described by the following equation.

$$2H_2O \ \rightarrow \ 2H_2 + O_2$$

In the next chapter we shall assess the current status of the global hydrogen industry before in subsequent chapters considering the trajectory for innovation for this fast-moving aspect of energy technology and policy.

References

Bulletin H (2022) Hydrogen colours codes [18 February 2022]. https://www.h2bulletin.com/knowledge/hydrogen-colours-codes/
California Air Resources Board (2021) An action plan for carbon capture and storage in California. California
California Energy Commission (2019) Joint Agency Staff Report on Assembly Bill 8: 2019 Annual Assessment of Time and Cost Needed to Attain 100 Hydrogen Refueling Stations in California
McKinsey & Company (2022) Decarbonizing the grid with 24/7 clean power purchase agreements. [cited 2022 20 June]. https://www.mckinsey.com/industries/electric-power-and-natural-gas/our-insights/decarbonizing-the-grid-with-24-7-clean-power-purchase-agreements

Department of Energy (2020) Hydrogen strategy: enabling a low carbon economy. Washington, DC

Department for Business (2022) E.a.I.S. British energy security strategy [cited 2022 7 April]. https://www.gov.uk/government/publications/british-energy-security-strategy/british-energy-security-strategy?module=inline&pgtype=article

Evans-Pritchard A (2020) Why fossil fuel dinosaurs may still have life in them yet. The Telegraph

ExxonMobil (2024) ExxonMobil and Shell selected to work with the Government of Singapore on a carbon capture and storage value chain. [cited 2024 5 June]. https://corporate.exxonmobil.com/locations/singapore/singapore-updates/news-releases/03012024_exxonmobil-and-shell-selected-to-work-with-the-govt-of-singapore-on-a-ccs-value-chain

S&P Global (2022) UK's gas grid ready for 20% hydrogen blend from 2023: network companies

Shell Global (2023) Hydrogen—what is it? Hydrogen Fuel & Projects | Shell Global. https://www.shell.com/energy-and-innovation/new-energies/hydrogen.html

Green Hydrogen Coalition (2022) HyDeal Los Angeles. [cited 2022 20 June]. https://www.ghcoalition.org/hydeal-la

Grid N (2022) The hydrogen colour spectrum [cited 2022 18 February 2022]. https://www.nationalgrid.com/stories/energy-explained/hydrogen-colour-spectrum

H21 (2019) H21 Leeds City Gate

hydeal (2022) Mass-scale green hydrogen hubs. [cited 2022 20 June]. https://www.hydeal.com/hydeal-ambition

HyNet (2018) HyNet North West: from vision to reality

International Energy Agency (2019) The future of hydrogen

Lawrence Livermore National Laboratory (2019) Energy flow charts [cited 2019 16 August]. https://flowcharts.llnl.gov/

Madhavi M, Nuttall WJ (2023) Modern global oil market under stress: system dynamics and scenarios. Proc Inst Civ Eng Energy 176(1):4–19

Martin C, Blumberg Z, Wason E (2021) Hydrogen's climate impacts, reimagining the social cost of carbon, and more [cited 2023 13 October]

Mittelsteadt C et al (2015) Chapter 11 - PEM electrolyzers and PEM regenerative fuel cells industrial view. In: Moseley PT, Garche J (eds) Electrochemical energy storage for renewable sources and grid balancing. Elsevier, Amsterdam, pp 159–181

Møller KT et al (2017) Hydrogen - A sustainable energy carrier. Prog Nat Sci Mater Int. 27(1):34–40

Newborough M, Cooley G (2020) Developments in the global hydrogen market: the spectrum of hydrogen colours. Fuel Cells Bull 11:16–22

Nuttall WJ, MacGregor I (2024) A Canadian case study of carbon dioxide removals and negative emission hydrogen production. ELSPublishing

Office of Tax Simplification (2022) Hybrid and distance working report: exploring the tax implications of changing working practices

Patonia A, Poudineh R (2022) Cost-competitive green hydrogen: how to lower the cost of electrolysers? Oxford Institute for Energy Studies

Politico (2023) EU reaches agreement on Russian oil products price cap. 2023-02-03. https://www.politico.eu/article/eu-reaches-agreement-on-russian-oil-products-price-cap/

Reuters (2023) India's Russian oil buying scales new highs in May. 2023–06–21. https://www.reuters.com/business/energy/indias-russian-oil-buying-scales-new-highs-may-trade-2023-06-21/

Reuters (2023) Startups race to strike hydrogen gold. 2023-09-07 [cited 2023 12 October]. https://www.reuters.com/business/energy/startups-race-strike-hydrogen-gold-2023-09-07/

Velazquez Abad A, Dodds PE (2020) Green hydrogen characterisation initiatives: definitions, standards, guarantees of origin, and challenges. Energy Policy 138:111300

YouTube (2020) Boris Johnson wants UK to be 'Saudi Arabia of wind power'. https://www.youtube.com/watch?v=79DDaPM9w0I

Chapter 2
Hydrogen Gains Momentum

Abstract If hydrogen is to fulfil its potential in the energy system then infrastructure will be important. This chapter introduces infrastructure considerations and points to the importance of pipeline systems and the regional context for innovation. The chapter also stresses the role of public policy in supporting bold hydrogen futures, whether based on fossil fuel conversion with carbon capture utilization and storage or renewable energy.

2.1 Introduction

In this chapter we summarise the current status of hydrogen in the energy system. We point to policy initiatives and business strategy considerations. We observe that, while the hydrogen economy is yet to fully break out of its current role as a foundation for today's fossil fuel economy (in fuels and fertilizers), there are clear signs that hydrogen is continuing to build momentum in the mid 2020s. A key consideration for the roll-out of the hydrogen economy is infrastructure.

2.2 Infrastructure

Hydrogen has been a commercial product with established supply chains and prices for more than 100 years. The established production methods and supply chains served the needs of major industrial users mainly in the petrochemical sector. In the USA, around 70% of hydrogen produced via SMR is used on the petroleum refining industry and 20% on fertiliser production (Office of Energy Efficiency & Renewable Energy 2018). Interest has grown enormously over recent years in the proposition that hydrogen represents a key enabler of a low-carbon high-technology future. Despite all the attention that has been given to hydrogen in the last few years, it is interesting how relatively invisible the existing hydrogen industrial capabilities have been for energy policy makers, academics and even parts of industry, until

© The Author(s) 2025
W. J. Nuttall et al., *Insights into the New Hydrogen Economy*,
https://doi.org/10.1007/978-3-031-71833-5_2

relatively recently. The change, however, is now well underway in both the UK and the USA. There is now a very high level of interest from industry and government in the potential to build up from the existing hydrogen industrial base. Such ideas are a theme discussed in *Fossil Fuel Hydrogen* (Nuttall and Bakenne 2020). Looking to the future, one can observe that a viable supply chain is a prerequisite for the emergence of a material hydrogen market. The good news is that many parts of the world already have developed hydrogen supply chains and the issue in those cases becomes one of scale-up and decarbonisation.

When considering the future of hydrogen, one soon concludes that infrastructure is a core consideration. By way of context, it is important to note that today, in the UK, far more energy originates in the form of chemical energy (i.e. fossil fuels) than as primary electricity (such as from renewable sources or nuclear power), see Fig. 2.1. It is clear that the proportion of inland energy arising from primary electricity will rise, but will that be sufficient for a near-complete electrification of the UK energy economy by 2050?

Concerning the prospects for hydrogen, its production needs supporting infrastructure: at the very least a reliable supply of either natural gas or clean electricity; it requires a means to supply its product to customers; and it needs to meet the needs of end users in heating, transport or other areas such as industrial applications (e.g. low carbon steel-making). Furthermore, whatever the energy system of the future, be it based upon a massive wave of electrification or based upon the roll out of hydrogen at scale one must remember that moving energy (either molecules or electrons) tends to be a bigger issue than actually producing it.

Green Hydrogen is expected to develop strongly in those territories with an abundance of renewable electricity over significant time periods. It is further favoured by regions with strong electricity infrastructures and a strong base of political support for low-carbon innovation. In contrast Blue Hydrogen is favoured in regions with a

Fig. 2.1 UK Inland Energy Consumption. RenewH2 2021 Source: UK Energy in Brief 2023 Open Government v3.0

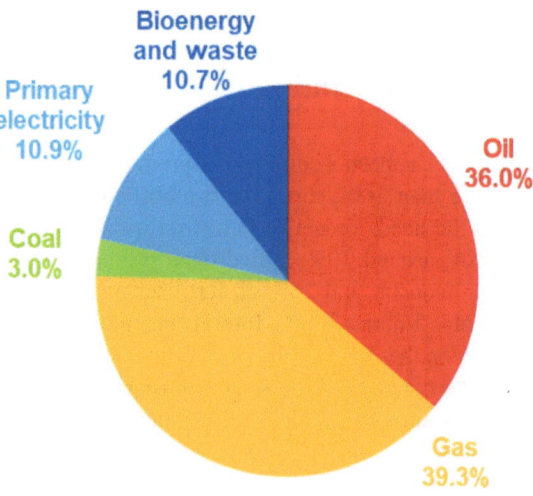

strong industrial base linked to chemical engineering, access to natural gas feedstocks and an ability to manage carbon capture utilisation and storage.

Hydrogen can be supplied to customers in a variety of forms, the suitability of which is primarily dependent on two factors; the rate at which it is supplied and the distances over which it is supplied (Nuttall and Bakenne 2020). For small scale local production, the most cost-effective method of transportation is using gas trucks. Over larger distances at all but the largest scales, it is preferable to liquefy hydrogen before transporting it by truck. At the largest scales and over modest to long distances, hydrogen is instead best supplied as a gas via pipeline.

2.2.1 Green Hydrogen and Infrastructure

Around the world industrial interest is growing quickly in Green Hydrogen production. A widespread perception, that Green Hydrogen has lower environmental impact than all other alternatives, drives enthusiasm amongst policy makers and in corporate boardrooms. Actually, one can debate the relative detailed merits, including environmental benefits, of Green Hydrogen over Blue. But it can be argued that both represent a significant improvement over current hydrogen production arrangements and, in this book, we esteem all approaches consistent with a rapidly delivered Net Zero economy. Perhaps the most important test to be applied to Green Hydrogen ideas and projects is the test of scale, which links to the question is scalability from where we are today. Are current proposals straightforward, or even achievable, by mid-century? We shall return to some of these ideas in Chapter 4.

2.2.2 Beyond Conventional Green Hydrogen

Conventionally Green Hydrogen imagines the use of renewable electricity (i.e. using wind or solar energy to split water in H_2 and oxygen), but there are also important renewable solutions based on biomass and biogases.

Two projects in North America are notable for both their scale and ambition:

- **RenewH2** will convert biogenic methane sourced from Wyoming into hydrogen via an SMR process (RenewH2 2021). The hydrogen will be commercialised via Hyzon a major promoter of FCEV commercial vehicles (Motors 2023). The arrangement between the two companies will help Hyzon to deploy a low carbon hydrogen fuelling infrastructure.
- In western Canada **H2 Naturally** (Hydrogen Naturally 2023) is developing a proposition that is described as being 'bright green'. The aim is to achieve truly negative carbon emissions by coupling hydrogen production from otherwise waste forestry products with bio-energy carbon capture and storage BECCS (Nuttall and MacGregor 2024).

2.2.3   Blue Hydrogen and Infrastructure

One key lesson arising from our knowledge exchange series of meetings has been that when contemplating infrastructures for Blue Hydrogen it is more important to consider the infrastructures relating to carbon dioxide management than it is to consider the needs for hydrogen itself. While hydrogen, of course, leads one to consider the role to be played by ports and pipelines it is usually more important to think of such infrastructures in terms of CO_2 handling. The fundamental logic here is that it is far easier to transport bulk CO_2 than it is to transport hydrogen at least in molecular form (i.e. as a compressed gas or cryogenic liquid). For instance, the existing 8500 km of pipelines in US can transport around 80 million tonnes of CO_2 per year. Looking ahead it has been observed that major pipeline infrastructure will be required by 2050 (over 13 times current size) in order to transport approximately 1361 million tonnes of CO_2 per year (Princeton University 2020).

Noting the importance allocated to CO_2 management in any Blue Hydrogen proposition, it is important to recognise that the issues not only relate to successful CO_2 transport, but also to its end-state. Initially industrial scale carbon capture and storage (CCS) was developed on the back of innovations associated with gas production and enhanced oil recovery (such as the Sleipner field in Norway (Petroleum 2022) and the Alberta Carbon Trunk Line in Canada (Major Projects Alberta 2022)) but increasingly CO_2 sequestration, also termed 'storage', has become a driving motivation in its own right. Furthermore, as the project learned from Ian Hibbitt at BOC-Linde at our first event, attention has shifted away from simply safe CO_2 disposal to include options for commercial use or 'utilisation'. One such example can be seen in The Netherlands where, in Rotterdam, CO_2 produced by Shell as a by-product of its steam methane reforming operations is processed by the industrial gases company Linde and others to achieve food-grade so that it can be used safely to enhance agricultural production via its OCAP (Organic CO_2 for Assimilation by Plants) network shown in Fig. 2.2 (taken from Ros et al. (2014)). The numbers associated with the Dutch activity are impressive: 500,000 Tonnes of CO_2 are supplied each year using a distribution pipeline system of more than 250 km connecting to 2500 hectares of growing space in greenhouses. The CO_2-rich atmosphere in the greenhouses enhances production of crops such as tomatoes. These and related measures also yield a significant reduction in agricultural energy use as growers shift away from natural gas heated greenhouses saving 140 million Nm^3 of natural gas each year, for further details see the OCAP Fact Sheet (OCAP 2020). (Note: Nm3 denotes 'Normal Cubic Metre'—that is a quantity of gas equivalent to the specified volume of gas, if at 1.01325 bar pressure and at 0°C temperature.)

The Open University knowledge exchange project heard from Naomi Boness that in the United States a closed-loop approach to synthetic fuels is developing. In such an approach natural gas is used to generate hydrogen, capturing CO_2 and then combining that CO_2 to create synthetic liquid fuels breaking the link to upstream oil supply. Meanwhile in the USA there is much interest in Direct Air Capture of CO_2 from the atmosphere, but this is a technology replete with technical challenges several

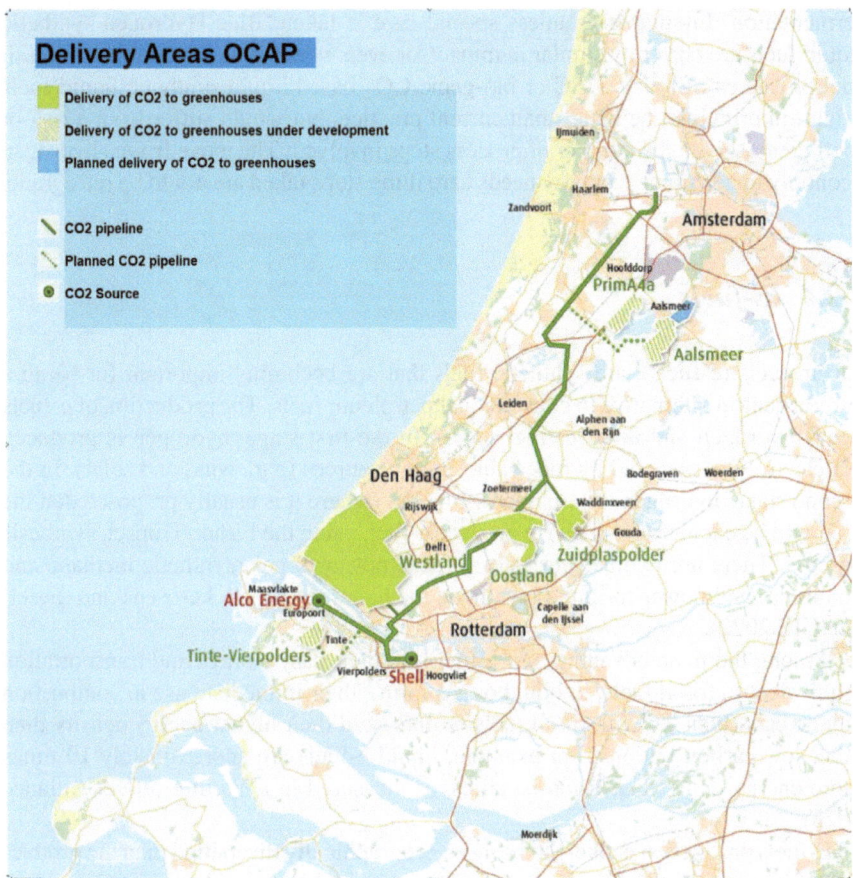

Fig. 2.2 The OCAP agricultural infrastructure led by Linde in The Netherlands. Source OCAP (CC-BY-ND) with English Translation of the key text – authors of this book are responsible for that translation

of which could prove to be a stumbling block. The existing DAC plants operate at small scale. There are 18 facilities in Canada, Europe and the USA and the first large scale DAC plant (approx. 1 MtCO$_2$/year) is expected to be operating by mid 2020s in the USA (International Energy Agency 2022).

Another consideration applicable to synthetic fuels is that if one were to make Blue Hydrogen from natural gas and then to take the captured CO$_2$ and later use it to make a synthetic liquid fossil fuel such as methanol or even synthetic diesel for combustion then no real benefit is seen. It is more energy intensive and more costly than the established processes of Gas to Liquids (GTL) already used to make synthetic fuels from natural gas. Indeed, Shell does this at commercial scale in Qatar at its Pearl GTL facility. The only potential benefit in the synthetic liquid fuels proposition is if the process uses CO$_2$ from DAC or bio-genic CO$_2$ such as that from bio-ethanol fermentation. In summary, unless special care is taken, Blue Hydrogen synthetic liquid fuels are currently similar in impact, or even worse, than the established fossil fuel processes of today. DAC or bio-genic CO_2 based liquid synthetic liquid fuels might, in principle, be better than current practice, but would suffer from a risk of inefficiency given the number of process steps involved. The move from a hydrogen economy to a synfuel economy needs care if the steps taken are not to be retrograde.

2.2.4   E-fuels

Electrofuels (e-fuels), are synthetic fuels that are becoming important for being a carbon neutral alternative to conventional petroleum fuels. The production of e-fuels needs hydrogen and involves two stages. In the first stage: hydrogen is produced via electrolysis powered by renewable energy sources (e.g.. wind and solar). In the second stage, hydrogen is combined with CO_2 (where it is usually proposed that the CO_2 will be captured from air) to produce e-fuel, using the Fisher-Tropsch synthesis process. There are two categories of e-fuels, power-to-gas (synthetic methane and ammonia) and power-to-liquid (synthetic methanol, crude oil, kerosene and diesel) (ENGIE 2022).

Among the main advantages of e-fuels are their easy storage and transportation (equivalent to fossil fuels, in liquid or gas form), their immediate use in combustion engine cars (blended with fossil fuels or pure) and their higher energy density than other low carbon options. For example, liquid e-fuels are approximately 10 times more energy dense than li-ion batteries, as measured in kWh/litre (eFuel Alliance 2023).

E-fuels production is likely to be more geographically diversified than the production of fossil fuels. Specific zones with strong renewable source have been identified, with some places being more efficient than others, see the Global power-to-X Atlas produced by Fraunhofer IEE in Kassel, Germany (Fraunhofer 2023).

A few initiatives of e-fuels production and testing are observed around the world. Porshe and the Highly Innovative Fuels (HIF) Global company are building a synthetic fuel plant in the south of Chile, with an expected production of 0.13mi l/year in phase 1 and up to around 55mi l/year by middle of the decade (Porsche Newsroom 2023). Aramco from Saudi Arabia and Stellantis have tested, and confirmed, that e-diesel and e-gasoline are compatible with 24 engines types, used by 28mi vehicles in Europe (Porsche Newsroom 2023).

However, there are some concerns that may jeopardise the scale up of e-fuels. The first one is related to their low end-to-end efficiency, estimated at 13% (in comparison to direct charging from battery EV with 77% and hydrogen fuel cell vehicle with 30%), increasing their production costs importantly (Porsche Newsroom 2023). The second one is in the regulation arena. Following a last-minute campaign promoted by Germany, in March 2023 the European Union agreed to exclude vehicles that run exclusively on e-fuels from the 2035 ban on new sales of combustion-engine cars (it was originally ruled without exceptions in the EU Green Deal). But more recently the European Union draft proposal of the new regulation is suggesting stricter rules on e-fuel cars (to be run on fully CO_2 neutral fuels), meaning that emissions along the value chain may be counted in addition to those from producing e-fuels. E-fuels community has expressed its concerns, finding nearly impossible to achieve a 100% reduction in emissions (Abnett 2023).

2.2.5   Hydrogen Moves Forward Worldwide

In Europe a very wide range of Green Hydrogen projects are being supported by subsidy and public policy. The north American examples given earlier in this chapter are chosen in part because they represent good examples of private sector-led thinking, albeit against a backdrop of public support. The intention in this book is to highlight areas of commercial leadership. The relatively large-scale ambitions in North America are to be commended. It remains to be seen where in the world leadership in Green Hydrogen at scale will emerge – the drivers of such development are, as with Blue Hydrogen, geographical. As with Blue Hydrogen all aspects of geography matter, physical, economic, political and social. We shall return to this theme at the end of this report.

Across the world, and despite examples of large-scale Green Hydrogen projects, generally the grandest ambitions for hydrogen energy roll-out have been seen in the domain of Blue Hydrogen. The U.S. Inflation Reduction Act (IRA) has however been a game changer for investment in Green Hydrogen. Electrolyser production sales are skyrocketing as a result of the U.S. IRA (@rechargenews 2022a, b). The act provides a production tax credit "45 V" of up to \$3/kg, depending on the "well-to-gate" CO_2 footprint of the hydrogen produced, as seen in Table 2.1 (taken from Satyapal (2023)). Green Hydrogen produced from wind or solar power, or Pink Hydrogen from nuclear power can readily achieve the maximum tax credit (Fig. 2.2). Blue Hydrogen via steam methane reforming and CCS achieves only the minimum tax credit. Methane emissions from wellhead to gate are yet another concern for Blue Hydrogen. Methane produced via fracking in the Permian Basin has a leakage rate more than 1% higher than the rate for natural gas produced from offshore Gulf of Mexico (Zhang et al. 2020). Allowed leakage rates for natural gas may be reduced by leakage of hydrogen itself. The logic behind such a measure is that leaked hydrogen reduces the rate of methane decomposition in the upper atmosphere. Consequently, Bertagni et al. (Bertagni et al. 2022) estimate a 9% leakage rate for hydrogen can be tolerated for a Green Hydrogen economy, but only 1% for a Blue hydrogen economy due to co-leakage of methane from the Blue Hydrogen supply chain. Uncertainty over ability to control methane emissions along the natural gas supply chain is an issue for Blue Hydrogen, which may require certification as "responsibly sourced gas". We shall return to the role and importance of gas leakages associated with Blue Hydrogen in Chapter 3 and again in Chapter 6.

As of 2022, there were 33 operational Green Hydrogen facilities across the North America region with a capacity of 691 thousand tons per annum (Nana Terra 2023).

Table 2.1 Clean hydrogen production tax credits. Source: Satyapal 2023 (US-DOE)

Carbon Intensity (kg CO$_2$ per kg H$_2$)	Max Tax Credit ($/kg H$_2$)
4–2.5	0.60
2.5–1.5	0.75
1.5–0.45	1.00
0.45–0	3.00

A number of Green Hydrogen facilities have been constructed in the US to serve local use for hydrogen powered forklifts (Plug Power) or ammonia plants (CF Industries), taking advantage of the ability to scale down electrolysis and not require a large, centralized facility. The CF Industries plant in Donaldsonville, LA was planned to employ a 20 MW alkaline electrolysis plant, and was scheduled for start-up in 2023, to make Green Ammonia. METI (The Ministry of Economy, Trade and Industry in Japan) looks to the U.S. Gulf Coast to decarbonize Japanese power and industry, using ammonia as a vector.

Figure 2.3 (taken from Department of Energy (2023)) shows planned and installed electrolyser capacity for the US in 2023 with a 500% increase over 2022. Much of this is driven by the IRA legislation

Recognizing the strength of the US Market due to the IRA, NEL announced plans for an automated gigawatt electrolyser manufacturing facility in Michigan (redactoramexico 2023).

Figure 2.4 (taken from Satyapal (2023)) shows the clean hydrogen production projects announced as of the end of 2022. Whilst it can be seen that some electrolysis with nuclear electricity and pyrolysis plant are present, hydrogen production is largely dominated by renewable electrolysis and methane reforming projects with carbon capture. Of these latter two, the split of projects is approximately equal.

A database of global hydrogen projects is maintained by The Hydrogen Council and McKinsey & Company where projects are categorised as either Announced,

Fig. 2.3 Planned and installed electrolyser capacity in the US for 2023. Source: US-DOE

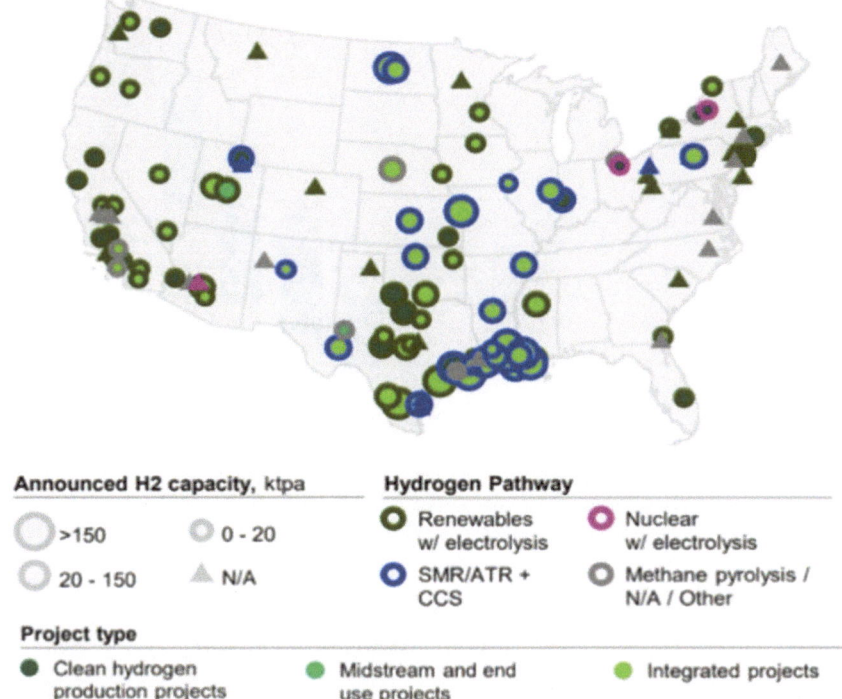

Fig. 2.4 Map of publicly announced clean hydrogen projects as of the end of 2022. Source: US-DOE

Planned, and Committed. While current capacity has more Blue Hydrogen than green, announced projects onstream by 2030 show a 2 to 1 ratio of green to blue projects, driven by policy and investment incentives for renewable energy, and as shown in Fig. 2.5 (taken from Hydrogen insights. (2023)).

2.3 The Hydrogen Business

A key focus of our interest has been hydrogen production at scale, in the terminology of the oil and gas sector – our concern has been for the emergence of hydrogen production as a 'material' consideration. There are many metrics against which materiality might be measured. The most obvious are by hydrogen's market energy share, or by its market revenue share.

Given hydrogen's diverse applicability, its market energy share is difficult to assess. Material hydrogen production in California is most likely to be in the form of transportation fuel, which already has some market penetration owing to significant infrastructure and policy support (California Energy Commission 2019). By contrast,

Cumulative production capacity announced, Mt p.a.

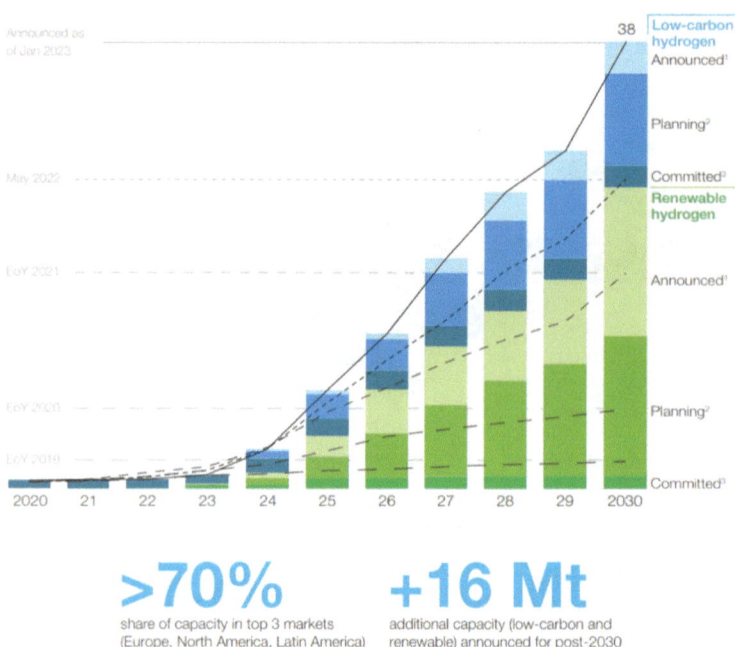

>70%   **+16 Mt**

share of capacity in top 3 markets   additional capacity (low-carbon and
(Europe, North America, Latin America)   renewable) announced for post-2030

Fig. 2.5 Clean hydrogen production volume by pathway. Source: Hydrogen Council with kind permission.

material hydrogen production in Texas is most likely to be in the form of feedstock for industry, for which there is already significant infrastructure and sales drive. In the UK and Europe, there is less of a pre-existing hydrogen market and penetration of hydrogen for transport or its supply via the natural gas grid are perhaps equally possible (Fuel Cells and Hydrogen 2 Joint Undertaking 2019; E4tech 2016).

Hydrogen's market revenue share is also difficult to assess, not only due to the diversity of its markets, but also due to the complex nature of revenue streams. A common discussion point in the literature examined as part of the background section above is the role of policy in nurturing a possible hydrogen economy. The prevailing wisdom seems to be that policy has both an enabling role, in the form of standards and regulation, as well as an incentivising role, in the form of subsidies and de-risking of investment.

As the International Energy Agency (International Energy Agency 2019) notes, hydrogen has seen many false starts. It is thus important to establish the criteria by which hydrogen production could be said to be material, not precluding the possibility that the hydrogen market might still collapse. One relevant consideration might be that material hydrogen production should be at the point where it no longer requires policy intervention to sustain its market. This points to a revenue-measured criterion, rather than an energy-measured one.

Perhaps one way of quantifying where such a threshold might occur would be using a basic financial model. Such a model is shown in Fig. 2.6. Given a choice of hydrogen production technology, the costs of its feedstocks and the attainable sale prices of both hydrogen and potential additional products, the profitability of a given proposition against the mass flow rate of production might be calculated. This would provide an understanding of threshold production rates, sale prices and feedstock costs required to make a given proposition viable.

Cost comparisons for proposed hydrogen production projects

Such tasks would be reliant on detailed analyses of the costs of proposed hydrogen projects. The basic financial modelling proposed above would provide a benchmark against which results might be tested. Provided a consistent framework and modelling approach were used, such modelling could be performed by university students as part of project work. Financial modelling should be performed to assess the economic performance of both Green and Blue Hydrogen, and their respective sensitivities to market pricing.

For Blue Hydrogen, it is vital to understand the sensitivity of cost to a carbon price. More specifically, understanding at what carbon price a shift to CCUS would become preferable to business-as-usual venting to the atmosphere. In addition, the sensitivity of a given proposition to the costs of its feedstocks, such as fossil fuels, oxygen, steam, or carbon dioxide, are similarly important. CO_2 costs would relate to unavoidable emissions, for example. These various costs are likely to have a significant impact on price of the hydrogen produced, and thus impact their viability

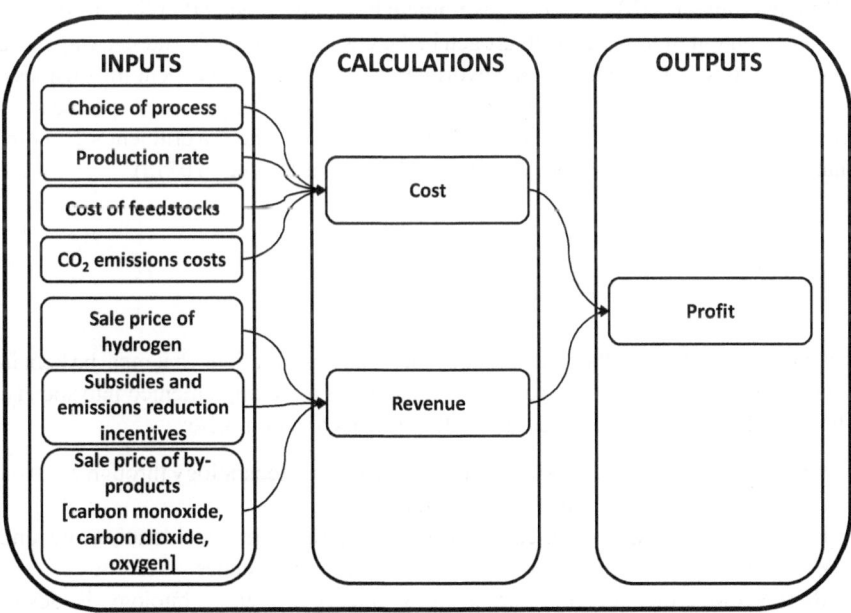

Fig. 2.6 Proposed financial benchmarking model. Image Source: Authors

in different geographical locations where, for example, these feedstocks might more easily be found.

For Green Hydrogen, it is necessary that the sensitivity of the economics of a given proposition to electricity price be understood. With this in mind, regions where the renewable electricity price might be sufficiently low to enable Green Hydrogen production should be identified. Existing studies show that Green Hydrogen is several times costlier than grey hydrogen, and around double the cost of Blue Hydrogen (Nuttall and Bakenne 2020). If these conclusions hold true, important insights might be gained by determining the sensitivity of Green Hydrogen propositions to the level of subsidy that might be available.

The parameter values at which Blue Hydrogen and Green Hydrogen reach cost parity should be a key point of interest going forward.

Employment

As part of his successful 2020 election campaign President Biden announced a plan for a clean energy revolution and environmental justice with an investment of $1.7 trillion. He claimed that his proposed policies would convert the US into a 100% clean-energy economy with net-zero greenhouse gas emissions. He further stressed that this initiative would create more than 10 million well paid jobs right across the United States (Harris 2022). Meanwhile in the UK, the then Prime Minister, Boris Johnson, set out a 10 Point Plan committed to removing 10 mega tonnes of carbon dioxide, generating 5GW of hydrogen by 2030 for industry (increased to 10 GW in April 2022), transport, power and homes, and aiming to develop the first town heated entirely by hydrogen by the end of the decade, creating 250,000 green jobs (HM Government 2020). The UK government has committed £60 million to support the development of low carbon hydrogen in the UK and to identify and scale-up more efficient solutions for making clean hydrogen from water (Department for Business, Energy and Industrial Strategy 2021b). About 9,000 hydrogen jobs are expected to be created by 2030 and up to 100,000 by 2050 as a result of the government's 10-point plan (Department for Business, Energy and Industrial Strategy 2021a).

2.4 Hydrogen Landscape

In a more holistic sense, there are broader societal considerations that must be kept in mind when considering how hydrogen production is to be implemented into society. These might rightly be considered enabling factors and comprise:

– **Policymaking** which can be an enabler for hydrogen technology through creation of subsidy and directing of research funding.
– **Industrial Strategy** which can be shaped by a range of considerations beyond government policy and beyond short-term costs and revenues.
– **Regulation** which is vital for ensuring that hydrogen technology is developed safely and sustainably and for enabling the international compatibility of technology.

- **Public acceptance and understanding** which should not be understated as a vital enabler for the adoption of hydrogen technology. These aspects link to political attitudes and hence have an impact on policy making and regulation.

Unlike the factors in Sect. 2.3, these are likely to have less of an effect on the route by which hydrogen is produced. On the other hand, their status as enablers mean that their handling will have a direct effect on the success of any potential hydrogen market. Whilst the consideration of all three is part of the scope of this book, it is in the making of policy recommendations where the most important output of this book probably will reside until industrial activity pulls ahead of government thinking Signs of that transition in thought leadership are already emerging, especially in North America (see Chapter 4).

2.4.1 Policymaking

Early adoption of technology, whether it be an FCEV or a hydrogen based steel refinery, is a risky venture. The risk to early adopters can be limited by policymaking that incentivises the adoption of hydrogen technology. This de-risking of investment will likely be as important to hydrogen production as it was, and still is, to renewable generators whose project risk is limited by access to guaranteed pricing and customers for their electricity. Ultimately, such policymaking is paid for by the public and thus it is important that in hydrogen's early days public money be directed towards markets with the greatest likelihood of success.

A less subtle issue is that of incentivising desirable hydrogen production. The easiest, most direct, and cheapest route to hydrogen production is grey hydrogen. Whilst this will greatly contribute to reducing CO_2 emissions from energy at its point-of-use, it will still produce significant amount of CO_2 emissions at source. It is likely that hydrogen production from fossil fuels will play some role in hydrogen production in any future hydrogen economy. It is possible that this will be transitory and simply bridge the gap as the industry shifts entirely to Green Hydrogen. More likely is that fossil fuels will be an important source of hydrogen for the foreseeable future. With this in mind, it is imperative that CCS technologies be incentivised, either through implementation of a sufficiently punitive carbon price for emitters or though subsidy of Blue Hydrogen technology, or via a combination of both. The continued dominance of Grey Hydrogen in the hydrogen economy must be brought to an end.

2.4.2 Industrial Leadership

When the history is written of humanity's response to the realisation that climate change is a serious problem, it seems possible that the response will be divided into two distinct thirty-year phases. In the first phase starting around 1990 governments around the world were the centres for concern and action and the climate emergency was seen as a policy problem. It seems increasingly likely that we are now in a second thirty-year phase in which actions by commercial entities will take leadership away from governments. While the first phase has made some impressive progress, for example in electricity system adjustments in the most developed economies, it is in the second phase, from 2020 to 2050, that the most dramatic progress is likely to be seen.

There are multiple drivers behind why corporate decision making is now getting ahead of government thinking. One example is the rising importance of Environmental, Social and Governance (ESG) goals in corporate strategy. Another related concern is the increasing level of corporate engagement with the need to reduce 'scope emissions'. Broadly these emissions are at three philosophical levels representing ever tighter levels of accountability and hence action. Scope emissions may be summarised:

- Scope 1 covers the direct emission of greenhouse gases from sources that are owned or controlled by the reporting organisation. For example, for an international oil and gas company this would include product transport emissions and methane leakage in the supply chain.
- Scope 2 covers indirect emissions from the generation of purchased electricity, steam, heating and cooling consumed by the reporting company. In our example this would include electricity purchased in support of refining operations.
- Scope 3 includes all other indirect emissions that occur in a company's value chain. For an international oil and gas company this would be the end-user emissions associated with the fossil-fuel product itself.

While there are logical issues to consider around the risk of double counting (in that one entity's scope 3 emissions can be another entity's scope 2 emissions) the pressure for corporate annual reports to consider scope emissions and to push to higher scope reporting and action is already driving significant change.

Another aspect in which corporate boardroom issues are translating into significant action on climate change relates to the criteria for executive remuneration. Once 'shareholder value' was the guiding principle of boardroom decision making, but increasingly corporate strategies have broadened to give significant weight to environmental harms. As the strategy has evolved, so have the criteria for executive remuneration. Senior executives are now explicitly incentivised to make dramatic moves to mitigate climate change impacts. These incentives are adding to range of drivers putting corporate action ahead of government action in the second phase and the push to 2050. The discussion meetings convened prior to the preparation of this book aimed to facilitate academia-industry dialogue for such a period in the hope of achieving an efficient and successful journey to Net-Zero.

We are pleased to note that for a growing range of industrial concerns the default 'business as usual' (BAU) scenario (i.e. the scenario to be considered before assessing the impact of a specific corporate decision) has changed in that it is now explicitly a journey to decarbonisation scenario. Previously the journey to decarbonisation scenario was regarded as an alternative to BAU – now increasingly it is BAU and that in itself is a major indicator that something profound has happened.

2.4.3 Regulation

Hydrogen is already an important industrial commodity and, as such, there is some existing understanding of how to handle, transport and store it. The use of hydrogen by the general public adds a further layer of complexity to this issue. Ultimately, however, whilst the approaches will necessarily differ, none of the requirements for these regulations are unprecedented. The use and handling of natural gas in homes is already rigorously regulated and only trained and experienced personnel are licenced to install natural gas burning equipment. The use of hydrogen in personal vehicles is a more complicated issue and one that will require significant research and public consultation.

It is not only domestic use that will require regulation. One of the most attractive features of hydrogen over electricity as an energy carrier is its ability to be transported over truly global supply chains. Many of the roadmaps and projects discussed in this document propose the import/export of hydrogen as a means to securing either hydrogen supply, or a hydrogen market. Such an international market will require internationally agreed standards and regulation on handling and storage systems. Those nations most willing to participate in the development of regulations will have an advantage in the manufacture of such systems.

2.4.4 Public Acceptance and Understanding

Public acceptance will likely be a major influencing factor as to where hydrogen might first find an eager market. In the example of the nuclear industry, we find a technology whose low-carbon credentials might, on paper, make it an important weapon in the fight against climate change. Despite this, lukewarm public and campaigner enthusiasm for the technology, as well as a number of high-profile accidents, have led to its abolition in several countries, most notably Germany.

Whilst hydrogen production is arguably far less controversial a technology, public acceptance plays perhaps an even greater role in it gaining a market foothold. FCEVs and displacement of natural gas in transmission networks are two frequently proposed uses of hydrogen. Both involve close proximity of the end-users to gaseous hydrogen, a flammable and potentially explosive material. Hydrogen use for FCEV applications does present safety concerns as does the alternative of BEV technology. There are numerous reports of fires associated with BEV vehicles – see, for example, a recent report from Paris (Bloomberg 2022). That particular example involves an innovative solid-state battery, a technology of much promise in the BEV world. The concerns of fire safety, however, also exist for more conventional BEV systems, but it must be emphasised that the petroleum fuelled ICE vehicles of today also present a significant fire risk.

In the UK the major chemicals company Ineos is a firm believer in the beneficial potential of hydrogen. In May 2024 the Chief Executive of Ineos Automotive Lynn Calder was reported to have said of the UK situation: *"We all understand the targets that we've got to achieve and take it very seriously. It's just about how we get there. And I think with a single horse that we're trying to back at the moment in the form of electrification in this country, and in Europe, I think we're going to fail"* (Auto Express 2023). It is reported that she went on to say *"Drivers are the cohort that has been completely left behind in this conversation [...] As someone with fresh eyes, you come and look at what the government in the UK and indeed the European Union has set out as a plan to decarbonisation. And that is on one technology platform, which is electrification for pretty much all use cases. I think it's gonna take a multiple solution, a multi powertrain strategy for this to work, and that is not where we sit today in either the UK or Europe"* (Auto Express 2023).

More generally in the British context, where there has been hope for a large-scale rollout of hydrogen via the decarbonisation of home heating, there are special aspects of public acceptance to consider. The issues here include the challenges of imposing change on citizens concerning their homes. Safety is also a key issue in this context. Homeowners and tenants need to be confident that they are safe and that the proposed changes will be affordable. At present there is much public policy interest in domestic electric heat pump technology, but the efficient use of such technologies relies on relatively tightly sealed homes. Many British homes are more than 50 years old and if a heat pump were to be fitted it would be like trying to operate a refrigerator with the door open. The familiar use of a boiler in the home, representing the modern equivalent of the hearth, is a socially-established option that a switch to hydrogen could maintain. Such familiarity can help with social acceptance, but that reality should not reduce regulatory concern for consumer safety.

2.5   Investor Confidence in Hydrogen

During 2021 hydrogen technologies garnered a lot of attention in the stock market (Financial Times 2023). For investors in the UK stock market, there are three prominent stocks directly involved in the hydrogen energy market. One of them, ITM Power designs and makes electrolysers, while the other two listed companies - AFC Energy and Ceres Power, make fuel cells. Between 2019 and early 2021, shares in these companies experienced astronomical growth, in some cases up to a tenfold increase. UK investors started to see a possible ubiquitous role for hydrogen in a future low-carbon economy. Hydrogen started to be seen by some as the missing piece of the puzzle needed to ensure that the UK does indeed become a net zero economy by 2050. The market started to appreciate that hydrogen can be used to generate electricity, store energy, modernize gas-based heating systems and to decarbonise various industries. There's was also a realisation that clean hydrogen can play a vital role in balancing the future electricity system because it can be created from renewable energy when intermittent energy is high, and it can create energy through fuel cells when that intermittent energy generation is low. This latter aspect associates the future of hydrogen with the widespread market, and indeed societal sense, that the future energy system will be based on electrification. Companies linked to the commercialisation of fuel cells and electrolysers were perceived to be key enablers of a smoothed electricity system based upon Green Hydrogen. The optimistic assessment of Green Hydrogen and its role in the electrified future was sufficient to drive a wave of interest in hydrogen related stocks.

In 2021 however the market enthusiasm started to ebb away. The shares in all three UK listed companies started to drift downwards. Meanwhile policy-makers thinking started to extend to Blue Hydrogen in part driven by the sense that the Blue Hydrogen challenge is larger in scope and hence the opportunities are greater. Attenuating that positive assessment, however, is the recognition that Blue Hydrogen is linked to the fossil fuel energy sector which is increasingly been questioned, and indeed even avoided, by some investment portfolios. It is noteworthy that numerous universities have recently moved to divest from fossil fuels, and this can perhaps be interpreted as a sign of aversion to supporting industry-led Blue Hydrogen innovation (Horton 2022).

According to Financial Times (2023); Share News 2023), the increase in valuation shown in Fig. 2.7 for the three companies, was driven by a wave of investment funds focused on environmental, social and governance (ESG) factors (Financial Times 2023). In addition, other reasons investors were paying more attention to hydrogen as a future energy carrier and hence buying hydrogen stocks included:

1. British-based energy companies, BP and Shell, were increasingly referencing clean hydrogen in their scenarios and business models, motivated by the global transition to lower-carbon energy.

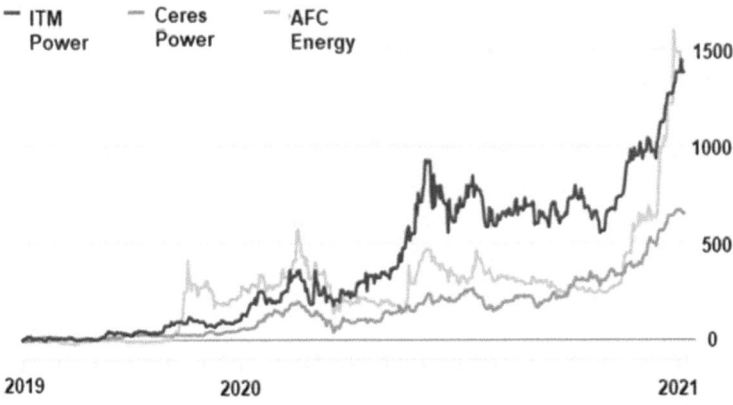

Fig. 2.7 Percentage change in share price of three key UK-based hydrogen companies rising to a peak in early 2021.Since then all three stocks have slowly fallen back. Image Source: Authors (after Financial Times). Data source: FactSet

2. Governments started to throw their weight behind hydrogen. For example, policy-makers in various countries planned, in detail, an agenda for a net zero carbon future. Governments started to plan post-COVID national industrial recovery plans referring to hydrogen and fuel cell innovations. By way of example, the UK government announced plans to spend around £400 m on hydrogen technology to achieve 10GW by 2030 and to attract up to £11b of private investment (Department for Energy Security and Net Zero 2023). Meanwhile Germany and the European Commission promised to spend 40–50 billion euros on a green energy support programmes, including to achieve over 40GW of electrolyte production capacity by 2030 (Investors Chronicle2023; Alpert 2023)

While it was interesting to see hydrogen-related stocks soaring, it was always the case that investments in the emerging hydrogen economy were regarded as risky and speculative (Forbes 2023). Generally such considerations associated with investor risk in the hydrogen economy remain very relevant today.

By way of example, we can refer back the three companies highlighted in Table 2.2. In the year to 30 April 2021 ITM Power reported only £3 million in sales and yet its market value touched £1.3 billion, representing a price-to-book (P/B) valuation ratio of 16.8 (Share News 2023). In the summer of 2021 AFC Energy hadn't yet generated any revenue, but showed a P/B ratio of 29.2. Ceres Power meanwhile had annual revenues of £23.5 million while achieving a lesser valuation, but it was still far from turning a profit. The consequence was a P/B ratio of 10.4 (Share News 2023). It should be stressed that the equivalent ratio for most companies yields a result between one and three. Such realties surely lay behind the market corrections that started to be seen after the early 2021 peak. By March 2022 the renewed Russian invasion attempt on Ukraine further upended the global energy business. In a turbulent world established fuels such as oil and natural gas soon attracted attention away from innovation and change.

Table 2.2 Global hydrogen stocks 12-months performance leading up to the peak of 2021

Global hydrogen stocks 12-month performance	
Company	%
FuelCell Energy	890
ITM Power (ITM:AIM)	713
Plug Power	566
AFC Energy (AFC:AIM)	544
McPhy	543
Powercell	352
Ballard	337
Ceres Power (CWR:AIM)	294
Bloom Energy	293
Nel	278

Data Source Google Finance, 14 August 2022

Larger technology companies, such as Johnson Matthey in the UK have established R&D to support development of hydrogen fuel cells and low carbon hydrogen projects (Share News 2023; Investors Chronicle 2023). More generally within the innovation landscape, there is a concern with the general pattern of smaller companies partnering with major companies (Bosch, Linde. Weichai Power and Acciona etc.) when scaling-up university spin-out and other emergent technologies to get them closer to commercialisation. The 2021 big rally in the sector prompted a realisation that a move away from fossil fuels would not just be about batteries, fuel cells will be needed as well as well. Despite the subsequent fall in share prices, we would posit that the logic of hydrogen for the future remains valid, although we acknowledge that many (including many investors) would disagree.

Looking beyond hydrogen production and directly associated engineering services as sources of industrial revenue, we note that, ITM Power and Ceres are focused on licensing out their valuable intellectual property rather than on manufacturing. The deduction is that the major companies will need access to this intellectual property, if hydrogen is to grow at scale. These companies have extensive R&D capabilities and resources, and consequently their involvement is also seen as key to de-risking the commercial development of hydrogen technology (Share News 2023).

We note that in 2021 Johnson Matthey terminated a substantial research and development activity dedicated to battery innovation. In so doing the company gave itself capacity to focus on other innovations, such as, for example, in hydrogen technology (BBC 2021). We note that the share price of this company has also seen downward pressure during 2022 and 2023.

We remind readers that nothing in this book comprises investment advice of any type. Indeed, our main message in this section has been to point to the need to be cautious, and to have a proper appreciation of risk when, investing in any energy technology or fuel. We offer no suggestion as to the future performance of any company named in this chapter, or indeed this book.

References

Abnett K (2023) EU set to demand e-fuel cars have no climate impact. 2023-09-22 [cited 2023 17 August]. https://www.reuters.com/sustainability/eu-set-demand-e-fuel-cars-have-no-climate-impact-document-2023-09-22/

Alpert B (2023) Hydrogen fuel-cell stocks are soaring. Yes, It's a Bubble

Auto Express (2023) Car pollution facts: from production to disposal, what impact do our cars have on the planet? [cited 2023 19 August]. https://www.autoexpress.co.uk/sustainability/358628/car-pollution-production-disposal-what-impact-do-our-cars-have-planet

BBC (2021) Johnson Matthey abandons electric car battery plans. [cited 2023 13 October]. https://www.bbc.com/news/business-59244991

Bertagni MB et al (2022) Risk of the hydrogen economy for atmospheric methane. Nat Commun 13(1):7706

Bloomberg UK (2022) Paris transit operator to investigate bollore electric bus fire [cited 2022 20 June]. https://www.bloomberg.com/news/articles/2022-04-05/paris-transit-operator-to-investigate-bollore-electric-bus-fire

California Energy Commission (2019) Joint Agency Staff Report on Assembly Bill 8: 2019 Annual Assessment of Time and Cost Needed to Attain 100 Hydrogen Refueling Stations in California.

Department for Business, Energy and Industrial Strategy (2021a) UK hydrogen strategy. HM Government

Department for Business, Energy and Industrial Strategy (2021b) £166 million cash injection for green technology and 60,000 UK jobs [cited 2022 20 June]. https://www.gov.uk/government/news/166-million-cash-injection-for-green-technology-and-60000-uk-jobs

Department of Energy (2023) Electrolyzer installations in the United States, in DOE Hydrogen Program Record.

Department for Energy Security and Net Zero (2023) Hydrogen investment roadmap: leading the way to net zero

E4tech (2016) Hydrogen and fuel cells: opportunities for growth - a roadmap for the UK

eFuel Alliance (2023) What are eFuels? [cited 2023 30 November]. https://www.efuel-alliance.eu/efuels/what-are-efuels

ENGIE (2022) E-fuels, what are they? [cited 2023 30 November]. https://www.engie.com/en/news/e-fuels-what-are-they

Forbes (2023) Why hydrogen stocks like fuelcell and bloom energy are lagging this year. [cited 2023 13 October]. https://www.forbes.com/sites/greatspeculations/2023/10/03/why-hydrogen-stocks-like-fuelcell-and-bloom-energy-are-lagging-this-year/

Fraunhofer IEE (2023) Global PtX Atlas. [cited 2023 30 November]. https://maps.iee.fraunhofer.de/ptx-atlas/

Financial Times (2023) Fuel-cell producers jump on new hydrogen 'hype cycle'. https://www.ft.com/content/ccbdd868-5499-11ea-90ad-25e377c0ee1f

Fuel Cells and Hydrogen 2 Joint Undertaking (2019) Hydrogen roadmap Europe

Harris B (2022) Climate: 10 million clean energy jobs [cited 2022 20 June]. https://joebiden.com/climate-labor-fact-sheet/

HM Government (2020) The ten point plan for a green industrial revolution

Horton H (2022) 100 UK universities pledge to divest from fossil fuels. 2022–10–27 [cited 13 November 2023]. https://www.theguardian.com/education/2022/oct/27/uk-universities-divest-fossil-fuels

Hydrogen Naturally (2023) Bright Green™ Hydrogen [cited 2023 15 September]. https://www.h2naturally.com

Hyzon Motors (2023) Zero emission, hydrogen-powered vehicles [cited 2023 1 January]. https://www.hyzonmotors.com/

International Energy Agency (2019) The future of hydrogen

International Energy Agency (2022) Direct air capture: a key technology for net zero.

Investors Chronicle (2023) Is now the time to invest in hydrogen? https://www.investorschronicle.co.uk/news/2021/01/06/is-now-the-time-to-invest-in-hydrogen/

Major Projects Alberta (2022) Alberta carbon trunk line [cited 2022 20 June]. https://majorprojects.alberta.ca/Details/Alberta-Carbon-Trunk-Line/622

Nana Terra (2023) 5 US Green hydrogen projects starting in 2023. https://www.airswift.com/blog/green-hydrogen-projects-usa

Norwegian Petroleum (2022) SLEIPNER ØST. [cited 2022 20 June]. https://www.norskpetroleum.no/en/facts/field/sleipner-ost/

Nuttall WJ, Bakenne AT (2020) Fossil fuel hydrogen. Springer, Bern

Nuttall WJ, MacGregor I (2024) A Canadian case study of carbon dioxide removals and negative emission hydrogen production. ELSPublishing

OCAP (2020) Pure CO_2 for greenhouses, B. Linde, Editor

Office of Energy Efficiency & Renewable Energy (2018) Fact of the Month May 2018: 10 million metric tons of hydrogen produced annually in the United States [cited 2022 20 June]. https://www.energy.gov/eere/fuelcells/fact-month-may-2018-10-million-metric-tons-hydrogen-produced-annually-united-states

Porsche Newsroom (2023) eFuels pilot plant in Chile officially opened [cited 2023 17 November]. https://newsroom.porsche.com/en/2022/company/porsche-highly-innovative-fuels-hif-opening-efuels-pilot-plant-haru-oni-chile-synthetic-fuels-30732.html

Princeton University (2020) Existing CO2 pipeline network in 2020. [cited 2022 20 June]. https://www.whitecase.com/sites/default/files/2021-01/carboncapture-infographics4.pdf

@rechargenews (2022a) ANALYSIS | Why the US climate bill may be the single most important moment in the history of green hydrogen. 2022–08–09. https://www.rechargenews.com/energy-transition/analysis-why-the-us-climate-bill-may-be-the-single-most-important-moment-in-the-history-of-green-hydrogen/2-1-1275143

@rechargenews (2022b) 'Game changer' | Shares in hydrogen-focused companies soar in wake of US climate bill's H2 tax credits. 2022–08–11. https://www.rechargenews.com/energy-transition/game-changer-shares-in-hydrogen-focused-companies-soar-in-wake-of-us-climate-bills-h2-tax-credits/2-1-1276425

redactoramexico (2023) Nel announced its plans to build a new automated gigawatt electrolyser manufacturing facility in Michigan - Hydrogen Central. 2023-05-04. https://hydrogen-central.com/nel-announced-plans-build-new-automated-gigawatt-electrolyser-manufacturing-facility-in-michigan/

RenewH2 (2021) Fuelling the future with renewable hydrogen. [cited 2023 1 January]. https://www.renew-h2.com/

Ros M et al (2014) Start of a CO2 hub in Rotterdam: connecting CCS and CCU. Energy Procedia. 63:2691–2701

Satyapal S (2023) U.S. DOE Hydrogen Program Annual Merit Review (AMR) Plenary Remarks

Shares News (2023) Don't get sucked into the hydrogen hype. https://www.sharesmagazine.co.uk/article/dont-get-sucked-into-the-hydrogen-hype

The Hydrogen Council and McKinsey & Company (2023) Hydrogen insights

Zhang Y et al (2020) Quantifying methane emissions from the largest oil-producing basin in the United States from space. Science Advances. 6(17): p. eaaz5120

Chapter 3
Hydrogen in the Near Term

Abstract This chapter focuses on the major near-term opportunities whereby hydrogen can contribute to modern energy policy goals. Key applications of hydrogen are described in areas such as industrial applications (both established and emerging), domestic heating, transport and mobility. Concerning the latter: ground, air and maritime opportunities are discussed. One aspect of transportation usage is highlighted – the use of hydrogen in fuel cell electric vehicles or in hydrogen combusting internal combustion engines. The industrial use of hydrogen is still dominated by the needs of the traditional petroleum industry together with demands from ammonia producers for agricultural fertilizer manufacture. Emerging industrial opportunities include low-carbon primary steel making.

3.1 Background

In this chapter we summarise and discuss the current and potential markets for hydrogen. The focus is on established processes and clearly visible business opportunities. The following characteristics for each market will be considered:

- The scale, level of maturity and potential for further expansion of current practices.
- Alternatives (competitors) to hydrogen in this area of the hydrogen market.
- Benefits and opportunities facing hydrogen-based technologies and their competitors.
- Barriers to the expansion of current approaches to decarbonised hydrogen production.

Subsequently, in Chapter 4 we will examine the status of hydrogen in various regions of the world. In Chapter 5 we consider various innovative ideas and developments that, in the authors' opinion, have the potential to advance significantly the prospects for low-carbon hydrogen supply and demand globally. The book will be concluded in Chapter 6.

Figure 3.1 illustrates an IEA assessment from June 2019 of the main sources and uses of hydrogen worldwide (International Energy Agency 2019). Ammonia and

W. J. Nuttall et al., *Insights into the New Hydrogen Economy*,
https://doi.org/10.1007/978-3-031-71833-5_3

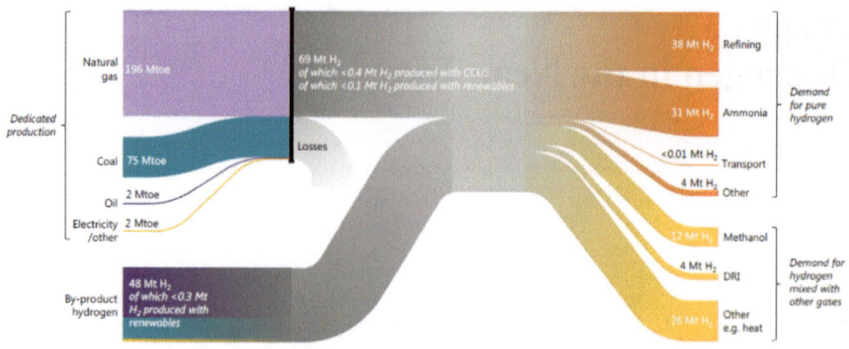

Fig. 3.1 Today's hydrogen value chains. Source: International Energy Agency (2019) International Energy Agency CC-BY-4.0

petrochemical conversion remain dominant aspects of hydrogen demand (International Energy Agency 2019). Methanol production and steel making also remain important uses of hydrogen at scale. In much of the developed world, and especially in the UK, domestic and industrial heating is presently dominated by natural gas, however, there already exists some instances where hydrogen is being explored as a blend into the natural gas grid, see discussion later in sections 3.2.4 and 4.2. There is significant potential for this to expand, with considerable current interest in Europe and the UK (H2 View 2020; H21 2019; HyNet 2018). Transport fuel currently remains a tiny market for hydrogen, but many regions consider this to be a key potential market of the future (California Energy Commission 2019; E4tech 2016; Verheul 2019; Hydrogen and Fuel Cell Strategy Council 2019; International Energy Agency 2015). Because much of this thinking is now well developed, we shall explore those issues in this chapter. More radical and innovative concepts in this area will be discussed in Chapter 5.

An important factor to be considered is the current cost of producing hydrogen. This can vary significantly depending on the methods of production, see Fig. 3.2 (taken from US The Federal Government (2020)). Hydrogen produced via solar PV electrolysis can vary enormously in cost, but nevertheless generally has been among the most expensive options (with an average cost over $12/kg according to the US DOE reporting in 2020) while hydrogen produced via coal gasification has been among the cheapest (at an average cost under $2/kg). For those parts of the world without effective carbon pricing, such cost disparities can greatly favour the production of hydrogen via environmentally harmful means. It is interesting to note in Fig. 3.2 that the addition of Carbon Capture and Storage (CCS) technology to fossil fuel hydrogen production does not represent a show-stopping economic problem (note for instance the relative cost of natural gas based SMR production with and without CCS and how the with-CCS cost compares to other low-carbon production options).

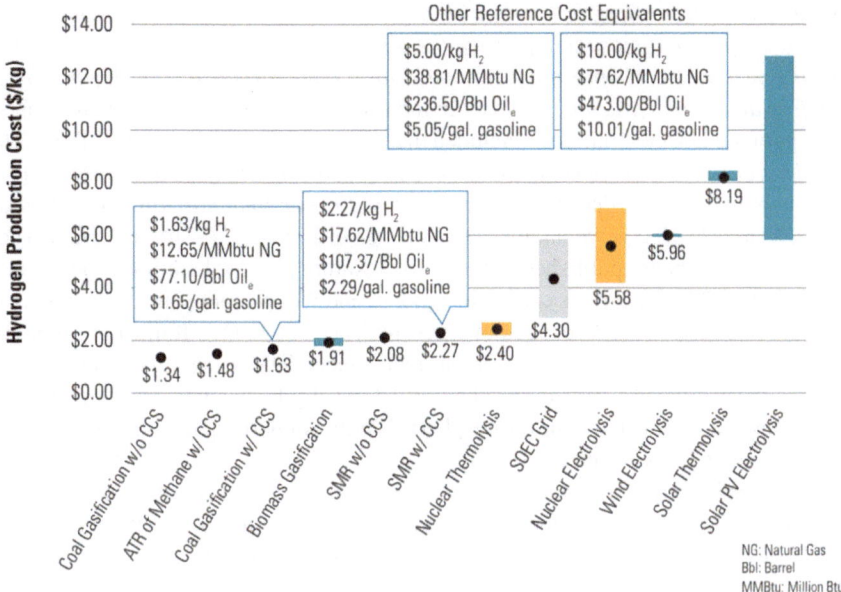

Fig. 3.2 Hydrogen Production Cost by Technology. The national hydrogen strategy 2020 Source: US-DOE, published July 2020

The data presented in Fig. 3.2 does not account for greenhouse gas leaks from distribution and industrial processing. Such emissions are known as fugitive emissions and they are a serious concern for Blue Hydrogen. Methane, the dominant component of natural gas, is a problematic greenhouse gas that on the timescale of 100 years or so, is far more potent than carbon dioxide. Any moves to the development of natural gas derived hydrogen (notably Blue Hydrogen) must consider such aspects and natural gas leakage must be reduced to the point of effective elimination. These requirements are challenging. Indeed, they are sufficiently challenging that some assert that Blue Hydrogen should have no enduring role in the Net Zero future. We authors remain optimistic that fugitive emissions can be greatly reduced, and we are pragmatic in taking the view that Blue Hydrogen has the potential to be a major enabler, not disabler, of a rapid global shift to a very low environmental impact energy economy.

Interestingly fugitive emissions are not just a concern for Blue Hydrogen, but rather are becoming a consideration for all sources of hydrogen including Green Hydrogen. Hydrogen is a simple diatomic molecule, and it is not a greenhouse gas. Concern has grown in recent years, however, that hydrogen can give rise to harmful indirect effects. Here we use the word 'indirect' in a somewhat different sense than in Chapters 1 and 2. Here we are referring to indirect atmospheric chemistry impacts, rather than impacts that are indirectly related to business activities. The idea of indirect emissions impacts from hydrogen leakage can be traced back to work by Dick Derwent and co-authors in 2006 (Derwent et al. 2006). Derwent et al.

posited: "*Because hydrogen reacts with tropospheric hydroxyl radicals, emissions of hydrogen to the atmosphere perturb the distributions of methane and ozone, the second and third most important greenhouse gases after carbon dioxide. Hydrogen is therefore an indirect greenhouse gas.*" In the summer of 2022 academic consultants writing for the UK government concluded that "*We estimate the hydrogen GWP(100) [that is over a 100-year period] to be 11 ± 5; a value more than 100% larger than previously published calculations*" and "*For a 20-year time horizon, we obtain a GWP(20) for H2 of 33, with an uncertainty range of 20 to 44.*" (Warwick 2022). The following year a paper published in the journal Nature provided similar conclusions (Sand et al. 2023). Such risks cannot be ignored and will prompt significant concern for fugitive emissions of hydrogen in a new hydrogen economy. It must be stressed however that these are estimates for indirect effects. The science is complicated, and the underlying assumptions are key. What exactly happens to hydrogen in the atmosphere? It is known that the atmospheric lifetime of hydrogen is short (at around 1.4 to 2.1 years (Pieterse et al. 2013)).

The global distribution of hydrogen in the atmosphere is unusual as noted by Derwent in a 2018 report for the UK government. He comments: "*Most trace gases with large man-made sources show higher concentrations in the northern hemisphere compared with the southern hemisphere, following the global population distribution. Hydrogen is unusual in having larger atmospheric sinks in the northern hemisphere compared with the southern hemisphere, following the global land surface distribution. The soil uptake sink for hydrogen causes higher concentrations in the southern hemisphere compared with the northern hemisphere, hence the inverted concentration distribution, because of the greater land surface area in the northern hemisphere.*"

We are not atmospheric chemists and cannot comment authoritatively on the lifecycle of a hydrogen molecule in the atmosphere, but the processes in play are clearly rich and complex. In Chapter 2 we pointed to such risks might be particularly important when hydrogen and methane might both leak from the same source point, such as a Blue Hydrogen production facility.

We take on trust the work of the experts, but we also note that even if things are as bad as is being suggested, the environmental case for hydrogen, especially Green Hydrogen, remains strong (Koch 2022). The emerging research suggests that those seeking to deploy Blue Hydrogen as a low-carbon solution must be extremely careful to eliminate leaks noting that technology's risks of both methane and hydrogen fugitive emissions. Hydrogen, once used, in either combustion or a fuel cell really is environmentally benign. Hence research is suggesting that hydrogen molecules must follow a path to end-use and they also must not be leaked to the atmosphere. Recent progress by the oil and gas sector towards low levels of fugitive emissions are therefore especially important if Blue Hydrogen is to be a credible option.

Fugitive leaks are not just an environmental concern, moves towards their elimination are a significant economic concern not captured by Fig. 3.2 (IEA 2023). Natural gas leaks and methane emissions should be very significantly reduced across the whole supply chain if Blue Hydrogen is to play a role as a complementary alternative to Green Hydrogen (Bauer et al. 2022). Innovation in the areas of both Blue and Green Hydrogen is progressing rapidly and hence it can be expected that clean hydrogen production costs will decrease overtime. The support of governments (both in terms of direct technology development and innovation incentives) is playing an important role on this downward trend. For example, in June 2021 the US Department of Energy launched the first Energy Earthshot - the Hydrogen Shot. It seeks to reduce the cost of clean hydrogen by 80% with a target of $1/Kg within a decade. This has become known as the goal of "1 1 1". The Hydrogen Shot involves the provision of funds to demonstration projects consistent with the goal (Office of Energy Efficiency & Renewable Energy 2021).

It is not the primary purpose of this book to provide introductory contextual information concerning hydrogen energy, rather it is the authors' intention to provide insight and information for those already familiar with current realities of, and potential for, hydrogen energy. As such, this book is not a hydrogen primer. For those seeking a more extended introduction to the role and prospects of fossil fuel-derived hydrogen we recommend the earlier book by Nuttall and Bakenne (Nuttall and Bakenne 2020). For those interested in knowing more about the history of hydrogen, we recommend Kendall et al. (Kendall et al. 2023). In this chapter we shall briefly review the current, and future, prospects for hydrogen, before in chapters 4, 5 and 6 providing information and insight that we hope will be of interest to the semi-expert reader.

3.2 Major Uses of Hydrogen Today

3.2.1 Ammonia Manufacture for Fertiliser

There are many scientific discoveries and technological innovations for which grand claims can be made concerning their role in shaping the twentieth century: the automobile, electricity, penicillin etc. One of the more important such innovations is amongst the least well known – the Haber–Bosch process for manufacturing synthetic ammonia from nitrogen in the air and hydrogen (usually produced from natural gas). Population growth in the nineteenth century had progressed on the basis of agriculture fertilised with unsustainable imports of guano (bird faeces) from the Chincha Islands off the coast of Peru. This rich source of natural fertiliser enabled a growth in nineteenth century agriculture sufficient to feed the rapidly growing and urbanising populations of Europe and North America. The story is told well by Chris Baker (Chemistry World 2023). In the twentieth century a new source of nitrogen-based fertiliser was needed. Synthetic chemistry would come to the rescue with the Haber–Bosch process, sometimes known simply as the Haber process after its scientific inventor, Fritz Haber. Carl Bosch industrialised the process.

Fritz Haber demonstrated his catalytic process in 1909 and for the innovation he was awarded the Nobel prize for chemistry in 1918, with the prize being presented in 1919 (The Nobel Prize 2023).

Today, in the twenty-first century the global population continues to grow, and synthetic fertilisers continue to be vital for food security. Statista reported in June 2023 that: *"total fertilizer consumption worldwide is forecasted to grow from 177.2 million metric tons in 2011/2012 to 191.8 million metric tons in 2022/2023, a 8.24 percent increase over the 11-year period. Nitrogen is the most used nutrient, followed by phosphate and potash"* (Statistica 2023a). Consequently, ammonia is a key industrial chemical with a global production approaching 150 million tonnes (Statistica 2023b). Almost all this ammonia is produced using the Haber-Bosch process and very large amounts of hydrogen feedstock and energy are required, see Table 3.1. Most of the ammonia produced globally is directed to the agricultural market (Statistica 2023b). Today the vast majority of the hydrogen needed to maintain global ammonia production is generated in the form of Grey Hydrogen (see chapter 1) (International Energy Agency 2019). Established production methods for synthetic ammonia suffer from severe environmental impacts (for both energy use and process feedstock reasons). Hence there is currently considerable interest in innovations to reduce environmental impacts, including the use of low-carbon production methods. It should be noted, however, that ammonia production is a very large-scale activity located at specific major industrial installations. Operations at scale provide for better process economics and easier site safety. This means that industrial ammonia production needs very large amounts of hydrogen (see discussion of captive versus merchant sources in Nuttall and Bakenne (Nuttall and Bakenne 2020). Such scale considerations and the existing industrial arrangements around natural gas supply are likely to favour shifts to Blue Hydrogen supply over Green Hydrogen alternatives, except in those territories where Green Hydrogen might be specifically incentivised by direct policy intervention.

3.2.2 Petrochemical Conversion and Upgrading

The petrochemical industry is the largest aggregate consumer of hydrogen today (International Energy Agency 2019). Hydrogen is primarily used by the petroleum industry for two purposes:

First for hydrocracking; the process of converting lower-quality viscous hydrocarbons into higher-quality more liquid molecules suitable for use as fuels. This links to the concept of "upgrading" whereby very viscous tars can be converted into liquids more akin to conventional crude oil. This has been a major industrial process in, for example, Alberta, Canada over several decades. The ongoing and significant increase in the extraction of poor-quality hydrocarbons in recent years is likely to continue in the near term and hence result in increased demand for hydrogen. The early 2020s war in Ukraine prompted by the Russian invasion of March 2022, led to sanctions on Russia and other measures highly disruptive to the global oil industry. These shifts are further likely to force international oil companies to consider lower quality resources, some of which may need significant hydrocracking (US Energy Information Administration 2022). Sanctions and geopolitics aside, the demand for

Table 3.1 Summary Table - Hydrogen for Ammonia Production

	AMMONIA	
Market information	**Demand from the hydrogen market:** 31 million tonnes annually (International Energy Agency 2019). This hydrogen is typically- produced locally via SMR and POX and is exclusively dedicated ammonia production **Market state:** highly mature **Potential for further expansion:** significant with the expansion of agriculture, even more so if ammonia is used as a medium for the transport of bulk hydrogen	
Pathway	**Dominant Pathway:** Haber–Bosch synthetic ammonia production, as discussed in main text	**Alternative pathways:** Examples: • Ammonia distillation from natural sources • Ammonia demand reduction, e.g. by substitution away from current farming practices based on synthetic fertilisers (note the growth in organic farming over the last 50 years)
Benefits	**Continuity with current farming practices:** Linking to: • Highly established networks • Almost total market dominance	**Appears to be more sustainable:** Linking to: • End consumer preferences • New public policy measures in key territories (especially agricultural policy)
Barriers	**Appears to be unsustainable:** • Requires continuous supply of substantial quantities of synthetic nitrogen-based fertilisers • Mismanagement of such chemicals can contribute to the risk of a perceived 'nitrogen crisis' as has been prominent in Dutch politics in recent years (Nuttall and Bakenne 2020)	**The challenge of Scale:** Some techniques (e.g. distillation) are rather dated (Dronsfield 2023). There is much scope for innovation towards more sustainable (e.g. bio-based) modes of ammonia and fertiliser production The key challenge will be to achieve production levels required even in a world shifting to more organic approaches to agriculture and hence in which fertiliser demand can be expected to fall. However, in the early 2020s the reality is that global fertiliser demand continues to grow, as noted in the main text

hydrogen in this industry will be driven for the international oil companies (IOCs) at least by the oil price as established in the form or Brent Crude or West Texas Intermediate indices. If the IOCs are unable to meet demand, the oil price will rise, and IOCs will be willing and able to pay more to process difficult molecules into saleable fuels. One obstacle to such innovation cycles, however, can be politics and the understandable public calls for 'windfall taxes' in scenarios where the IOCs appear to be generating super-profits based on nothing more than external events.

Second the petroleum industry uses significant amounts in another aspect of its refining operations – hydrotreating. Many territories around the world now ban the retail sale of gasoline and diesel fuels high in sulphur content. Sulphur is a natural component of many crude oils around the world and its removal (via the industrial processing of the sulphur to hydrogen sulphide) requires very large amounts of hydrogen.

Taken together, hydrocracking and hydrotreating ensure that the petroleum industry is the world's largest aggregate user of hydrogen (Table 3.2).

Table 3.2 Summary Table - Hydrogen and Petrochemical Industry

	Petrochemical industry	
Market information	**Demand from the hydrogen market:** 38 million tonnes annually (International Energy Agency 2019) **Market state:** highly mature **Potential for further expansion:** appears high given increased extraction of low-quality hydrocarbons, but future reductions in consumption driven by global decarbonisation efforts could erode demand	
Pathway	**Hydrocracking Dominant Pathway:** Hydrogen-based Today this is primarily used for producing JET-A aviation fuel (largely kerosene) and diesel (for road vehicle use) (US Energy Information Administration 2023)	**Alternative: Pathways:** There is the potential for other cracking techniques such as thermal cracking, steam cracking and fluid catalytic cracking (Nuttall and Bakenne 2020)
Benefits	**Conventional hydrocracking benefits from:** • Highly established networks • Almost total market dominance • Significant potential for expansion given the increase in extraction of lower-quality hydrocarbons	**Alternative approaches:** • Have minimal need for hydrogen supply and hence avoid supply-side risks • Have been demonstrated to be effective for a wide range of refined products, e.g. • Thermal cracking produces distillates, burner fuel and petroleum coke • Fluid catalytic cracking has a major role in the production of gasoline and liquefied petroleum gas • Steam cracking has the potential, through innovation, to displace established hydrocracking methods for some applications
Barriers	**Conventional hydrocracking:** Requires continuous supply of substantial quantities of hydrogen • This market is likely eventually to reduce if fossil fuel consumption falls in line with broader decarbonisation efforts • Consumption is likely to be driven by exogenous considerations, such as the oil price	**Alternative approaches:** • Significant energy needs for plant operations • The push for innovation may be diminished if the oil refining sector as a whole, is in retreat

3.2.3 Near-Term Uses – Clean Steel Making

Hydrogen can be used as an alternative to coking coal in the reduction of iron ore in steel production. This market, whilst considerably smaller than other industrial uses today, presents perhaps the greatest opportunity for future expansion. This is due to the ubiquity of steelmaking and the considerable contributions it makes to global carbon emissions, as well as the fact that hydrogen remains one of the only mature options for the decarbonisation of primary steelmaking. The use of electric arc technology for steel recycling is a different proposition. While important, it cannot alone meet global steel demand in total, nor in speciality areas.

Alternative pathways to low-carbon primary steel production include the combination of current approaches to steel making with retro-fit carbon capture and storage (CCS). An additional tactic, perhaps in combination with CCS, could be to use bio-based coke as a move towards a lower carbon steel industry. Hydrogen however, given its properties as a reducing agent as well as a fuel, remains a most interesting option for low-carbon steel manufacture (Table 3.3).

In February 2024, Nippon Steel, the largest steel maker in Japan announced that it had demonstrated a 33% greenhouse gas emissions reduction in blast furnace operations. The focus of the small scale test was to maximize substitution of coking coal with hydrogen. This is reported as being the highest level of emissions reduction achieved through hydrogen substitution (Steel Times International 2024). The company seeks to reduce blast furnace emissions further, so as to achieve an overall 50% emissions reduction.

3.2.4 Near-Term Uses – Hydrogen for Heating

In Europe, heating is the main use of energy by households with a share of around 64% of final energy consumption in the residential sector (Eurostat 2023). Hydrogen has long been proposed to decarbonise heating requirements which today are, for the most part, fuelled by natural gas. The UK, in particular, has a significant interest in this, with two projects proposing conversion of large domestic and industrial grids to full or partial hydrogen supply (H21 2019; HyNet 2018) and one project proposing the conversion of the UK's largest chemicals park to hydrogen (Equinor 2020). The scale of decarbonisation required in heating, the opportunities for phased rollouts, and the difficulties presented to electrification by many end users could represent the single greatest opportunity for hydrogen (Table 3.4). We shall consider this to alter in section 4.2.

Table 3.3 Summary Table - Hydrogen and the steelmaking industry

	Steel making	
Market information	**Demand from the hydrogen market:** 4 million tonnes annually (International Energy Agency 2019) **Market state:** nascent **Potential for further expansion:** very high given the significant contributions steelmaking using coking coal makes to carbon emissions globally and the need to decarbonise the sector	
Pathway	**Hydrogen:** Used as a heat source and reducing agent in place of coking coal (Nuttall and Bakenne 2020)	**Alternatives:** e.g. CCS, biofuel derived coke and electric arc furnace steel recycling
Benefits	• The only major industrially demonstrated option for decarbonisation of primary steelmaking • A potentially massive market given the importance of steelmaking, but this demand could be met in a rapidly expanded hydrogen economy	• Potential for retrofit of CCS to existing plants allowing current assets to continue • Some alternative approaches in some contexts appear to be lower cost than the hydrogen pathway, but some will struggle to achieve scale to meet demand (e.g. in particular steel recycling)
Barriers	• Requires industrial quantities of hydrogen ideally supplied as part of an industrial (multi-sector) cluster • Steelmaking is already established in specific geographies. Hydrogen production is likely to emerge at scale in regions previously focussed on oil and gas operations. The challenge in a cluster approach is for hydrogen production and supply to be geographically compatible • Increases steelmaking costs	• CCS remains difficult to perform on highly contaminated carbon dioxide waste streams such as those produced by steelmaking in comparison with those from coal-fired generation (McKinsey & Company 2020; U.S. Energy Information Administration 2021) • Electric Arc Recycling requires abundant low carbon electricity and access to large amounts of high-quality waste steel. Again, geographical coincidences are required

3.2.5 Near-Term Uses - Transport and Mobility

Hydrogen has long been touted as a means for decarbonisation of transport and it is here that it faces perhaps its largest challenges and its greatest opportunities. So far, at least, the long-expected breakthrough into large-scale hydrogen use has failed to materialise.

We introduced the possibilities for hydrogen use for and transport and mobility in Chapter 1. Figure 3.3 summarises the four main areas of potential transport use.

As a sector, transport and mobility includes a wide range of disparate activities and technologies. We can divide these challenges into road transport and other applications (rail, shipping and aviation) and we shall consider all these various aspects in the next section. First, however, we shall present a summary table, Table 3.5,

Table 3.4   Summary table - Hydrogen for domestic heating

	Heating	
Market information	**Demand from the hydrogen market:** some market presence - generally at the level of test and demonstration projects (see chapter 4) **Market state:** nascent **Potential for further expansion:** enormous considering the difficulty of decarbonising home heating	
Pathway	**Hydrogen:** Either as a blend with natural gas (for emissions reduction) or as a 100% replacement for natural gas	**Alternatives:** electric heating – especially air-source heat pumps, and where possible, ground source heat pumps
Benefits	• Can approach 20% blend in existing gas networks (in the UK at least) with no change to consumer equipment • Significant changes have been made to the natural gas network including polymer pipework better suited to hydrogen transmission than steel • Potential to link to major opportunities in the industrial heating sector • Phased rollouts, both regionally and in hydrogen concentration are possible	• Requires no major changes to existing electricity distribution infrastructure • Technology for electric heating already exists and is in widespread use • Can, to some extent, be aligned to fluctuations in renewable electricity generation via smart grid technology
Barriers	• Full 100% hydrogen conversion is likely to require significant changes to both infrastructure and consumer equipment • The necessary technology has yet to be proven commercially viable • The hydrogen supplied will not be pure partly because of the need to add an odorant for safety reasons, just as with natural gas • Public attitudes are not generally supportive. Safety concerns are frequently mentioned	• Energy requirements are substantial. Heating accounts for around a third of primary energy use which would require a substantial increase in electricity generation • Electric heat requires good home insulation. Many British homes are old and poorly insulated • Air source heat pumps are bulky • Air source heat pumps are expensive capital items, although they are subsidised in many territories, including the UK • Electric heating may well prove unsuitable for industrial high-grade heat requirements, so beneficial synergies with industrial developments are less likely

Fig. 3.3 Potential transport
applications for hydrogen.
Grey font denotes potential
hydrogen gas turbine. With
thanks to Marc Cochrane

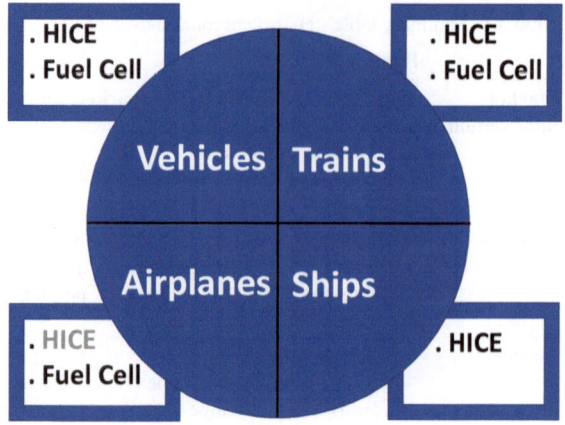

(focussed on road transportation uses) to sit alongside those presented earlier in this
section.

Building on considerations raised earlier around the importance of geography
in determining how hydrogen supply and demand will evolve, in Chapter 4 we
shall consider hydrogen innovations and developments in various key territories
around the world. In that context we shall return to consideration of UK hydrogen

Table 3.5 Summary table - Hydrogen for transportation

	Transport	
Market information	**Demand from the hydrogen market:** Generally, pilots and demonstration projects only **Market state:** nascent **Potential for further expansion:** very high, but major and established competition from BEVs	
Pathway	**Hydrogen**	**Alternatives: battery electric vehicles, biofuels**
Benefits	• Short refuelling times on par with fossil fuel engines • Greater suitability for haulage, mass transit and potentially aviation and shipping • Some rollouts of hydrogen trains • Some demonstrated markets, particularly California	• Established manufacturers in BEV personal mobility • Established electricity infrastructure • Rollout of fast charging is already taking place
Barriers	• Lack of hydrogen supply infrastructure • Potentially a short window of opportunity for deployment	• Low suitability for haulage, aviation, and shipping • Long charge times • Potential suitability issues for urban areas with minimal access to charging

heating ambitions. In Chapter 5, with help from guest contributors, we shall highlight important issues for hydrogen beyond those aspects that are already widely known.

We have highlighted hydrogen for heating as an opportunity that links closely to issues of industrial clusters and gas distribution networks. As such its prospects are heavily determined by geography in all its forms: physical, political, social, economic and industrial – to name but a few. We shall return to such issues in Chapter 4.

Hydrogen for transport and mobility is a major potential application area for hydrogen and we shall devote the rest of this chapter to considering such issues in greater detail.

3.3 Focus: Hydrogen for Mobility

3.3.1 Road Transport

At the outset it must be said that both the use of hydrogen as a fuel in either an ICE or in fuel cells where it is used to generate electricity to drive electric motors has been demonstrated to be feasible and hence could, in principle, replace traditional hydrocarbon combustion engines. The use of hydrogen as a motive fuel eliminates tailpipe emissions of harmful CO_2. In the case of FCEV powertrains, the only emission will be harmless water. In the case of Hydrogen fuelled ICE systems there is the possibility of the production of harmful nitrogen oxides (NOx) emissions (McKinsey & Company 2023). Such emissions are analogous to one of the key concerns associated with diesel fuelled ICE power trains in urban contexts. The issues of NOx relate in part to combustion control. This is an aspect of ICE design that is especially important for hydrogen fuelled ICE technology. It not only relates to emissions control but also to engine wear and hence reliability. It is sometimes argued that, with care, Hydrogen ICE design and operation can largely avoid NOx emissions (McKinsey & Company 2023) but even if this turns out not to be the case, exhaust scrubbing technologies can be incorporated, just as are used in today's diesel vehicles.

Fundamentally Hydrogen FCEV vehicles are electric vehicles. As such they have more in common with recently developed BEV technologies than with the traditional ICE approaches to transport and mobility. Technologies such as electric motors, power management systems, regenerative braking and interfacing to smart driver assist technologies are all common to FCEV and BEV technologies. Of course, battery development is a key aspect of BEV technology, but even that can be of direct benefit to FCEV developers. It seems highly likely that any future development of FCEV technology may, in fact, be a hybrid development between FCEV and BEV approaches. That is to say, every FCEV vehicle will also have a battery for energy storage and ancillary services reasons. Batteries installed in FCEV have a different role in comparison with those from BEV (they don't need them because they produce their own electricity). The implementation of plug-in capabilities in FCEV to charge the batteries could be an option however currently FCEV batteries are mainly for

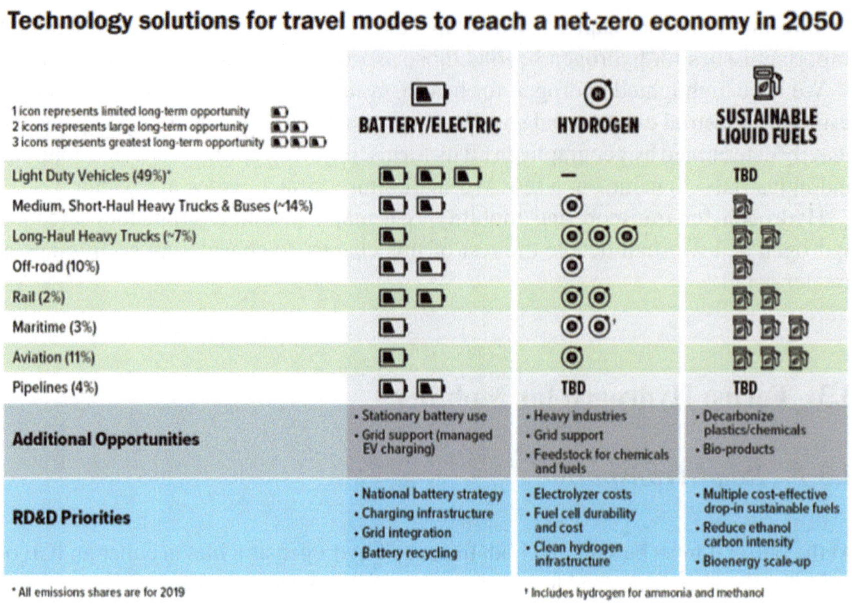

Technology solutions for travel modes to reach a net-zero economy in 2050

	BATTERY/ELECTRIC	HYDROGEN	SUSTAINABLE LIQUID FUELS
1 icon represents limited long-term opportunity ▪ 2 icons represents large long-term opportunity ▪▪ 3 icons represents greatest long-term opportunity ▪▪▪			
Light Duty Vehicles (49%)*	▪ ▪ ▪	—	TBD
Medium, Short-Haul Heavy Trucks & Buses (~14%)	▪ ▪	◎	⛽
Long-Haul Heavy Trucks (~7%)	▪	◎ ◎ ◎	⛽ ⛽
Off-road (10%)	▪ ▪	◎	⛽
Rail (2%)	▪ ▪	◎ ◎	⛽ ⛽
Maritime (3%)	▪	◎ ◎'	⛽ ⛽ ⛽
Aviation (11%)	▪	◎	⛽ ⛽ ⛽
Pipelines (4%)	▪ ▪	TBD	TBD
Additional Opportunities	• Stationary battery use • Grid support (managed EV charging)	• Heavy industries • Grid support • Feedstock for chemicals and fuels	• Decarbonize plastics/chemicals • Bio-products
RD&D Priorities	• National battery strategy • Charging infrastructure • Grid integration • Battery recycling	• Electrolyzer costs • Fuel cell durability and cost • Clean hydrogen infrastructure	• Multiple cost-effective drop-in sustainable fuels • Reduce ethanol carbon intensity • Bioenergy scale-up

* All emissions shares are for 2019   † Includes hydrogen for ammonia and methanol

Fig. 3.4 Summary of travel mode attributes in a low carbon economy. Source: (Department of Energy (2023) The U.S. National Blueprint for Transportation Decarbonization 2023)

recapturing breaking energy, providing additional power during short acceleration events and to smooth out the power delivered from the fuel cells (Alternative Fuels Data Center 2023). It is worth remembering that every ICE vehicle today has a battery and an onboard electrical system.

It is conventionally argued that a standalone hydrogen FCEV powertrain will have advantage over a standalone BEV powertrain in the largest and heaviest vehicles travelling the longest distances, as shown in Fig. 3.4. BEVs are largely unsuitable for heavy haulage which will require the energy density (kWh/kg) and rapid fuelling times (typically less than 15 min for a heavy truck). Hydrogen has the potential to meet these needs better than BEV alternatives (The Fuel Cell and Hydrogen Energy Association). Keeping such idea in mind, there can be expected to be faster development of FCEV technologies for trains, trucks and buses than for cars and motorcycles. Despite such conventional wisdom, let's start by considering hydrogen cars.

3.3.2 Hydrogen Cars

While, as we shall see later, the value proposition for hydrogen fuelled heavy vehicles, such as trucks, is fundamentally economic in nature the case for hydrogen cars is rather different. Key considerations for individuals considering a new car can include where they will park their car overnight and how this relates to their plans to refuel or recharge the vehicle. Many people living in city centre apartments or terraced housing struggle to access affordable BEV charging infrastructures. For such people very short-term smartphone app based car rental services might be good solution and indeed such realities might render new car leasing or purchasing unnecessary. With that said, however, many people who have grown used to their own car will want to maintain that freedom and convenience to which they have become accustomed. Such people might find that a hydrogen car represents a low-carbon continuation of habits and freedoms that they appreciate and value. These freedoms can be very personal, such as the freedom to smoke or vape, something one cannot usually do in a rental car. We admit the safety-related irony implicit in the suggestion that hydrogen cars might appeal to smokers. More importantly, however, one's choice of car, or indeed whether to have a car at all, is not something one can properly understand through simply a cost per mile analysis. If such considerations alone determined consumer choices, then many premium sports car brands simply would not exist.

We have attempted to explain that even in the absence of a clear economic case, hydrogen fuelled cars may find a role, but we must acknowledge that hydrogen fuelled technologies will struggle to enter the passenger car market owing to the rapidly growing presence of BEV brands. Many European automakers are making a bold shift towards BEV technology, notable among these is Volkswagen, following the embarrassment of the "dieselgate" scandal of 2015 (BBC 2015). As we shall discuss below, however, some large automakers – especially Toyota remain more sceptical about the global importance of BEV approaches to car technology. It must be acknowledged that so far hydrogen cars have generally struggled to get a foothold in the market. The Californian car market is one place that has made a strong attempt (California Energy Commission 2019), and over the years Germany and Japan have signalled firm interest in the future of hydrogen transport (Hydrogen and Fuel Cell Strategy Council 2019; Robiniusa 2018). In Germany, enthusiasm for hydrogen is nearly always firmly linked to an implied promise that the hydrogen involved will be green.

We shall consider below the various efforts by car manufacturers to make hydrogen cars a reality. Most of these projects have led to demonstration vehicles or small fleets prepared for specific customers. Relatively few have been cars that ordinary individual consumers can simply order from their local car dealership. The best example of a hydrogen car available for purchase has been the Toyota Mirai.

3.3.2.1 Hydrogen Car Companies

In this section we consider some prominent global car companies with a clear interest in the possibility of a future featuring hydrogen-powered cars.

While Volkswagen is making a firm bet on BEV technology, another German car-maker, BMW, has for a very long time been interested in the possibility of hydrogen powered cars. In the mid 2000s BMW explored the possibility of a hydrogen ICE car producing the Hydrogen 7 demonstrator. The car was never put into full-scale production. More recently BMW has returned to hydrogen developing the BMW iX5 – a variant of the company's long-established mid-sized luxury X5 SUV. In essence the iX5 is a concept car designed to demonstrate the viability of an H2 FCEV drivetrain (BMW 2023a). BMW correctly pitches the concept as an EV with a twist. The twist being that while one has essentially all the environmental and other benefits of a BEV one does not have the drawbacks associated with battery technologies and their charging requirements (T3 2023). BMW has built a demonstration "pilot fleet" of 100 vehicles that it is taking to several key markets around the world. In late summer 2023 the BMW pilot fleet came to the UK to help raise awareness of the potential for the hydrogen car generally and their own innovative role in this space (BMW 2023b).

The fuel cell module at the heart of the BMW iX5 is supplied to the company by Toyota as part of a collaboration between the two companies established in 2013 (T3 2023). Toyota is the pioneering developer of the Mirai, first unveiled in late 2014. The company is reported to have achieved global sales rising from 1,770 units in 2020 to 5,918 in 2021 (@rechargenews 2022). While the growth in numbers may appear impressive it was a consequence of deep discounting in the main FCEV car market – California. Generally global sales are measured in the thousands rather than the hundred of thousands, or millions, of units that a company like Toyota might expect for a commercial product. That said, the Toyota Mirai is a car you can buy, if you are fortunate enough to live in a supported market.

In September 2023 Toyota announced that it will produce a hydrogen FCEV version of its popular Hilux pick-up truck at its Burnaston plant in Derby, England (Crouch 2023). Toyota has long been a proponent of FCEV technology building upon its early great success in BEV hybrid technology (starting with the Toyota Prius launched in 1977). Toyota has however always been somewhat cautious about pure BEV technology. In July 2023, however, Toyota announced significant progress in its work on solid state batteries. These batteries contain no liquids unlike the batteries powering BEVs today. Toyota is reported to have claimed that its new batteries will have a range of 1,200 km and even more importantly be fully charged in 10 min (Davies 2023). The reported breakthrough concerns production engineering. The claimed performance yields the potential for future Toyota BEVs to avoid several key aspects of consumer resistance around range anxiety and slow charging. Toyota are reported to claim that their new batteries will be significantly smaller, lighter and less expensive than the current technology. Through such innovation Toyota may have found a way past its own corporate hesitation concerning the future of battery electric cars. Furthermore, if such next generation battery technologies can be brought to market, this could be sufficient to end any future prospects for hydrogen FCEV personal mobility.

- **GM and Honda Collaboration**

Meanwhile Toyota's Japanese rival Honda is reported to have entered into a collaboration with American automaker GM to produce a lower cost next generation fuel cell (Reuters 2023).The focus appears to be on producing a competitor technology to battery electric trucks rather than cars, at least initially. Honda is remembered as

Fig. 3.5 Dr. Bakenne standing next to AA Hyundai Nexo during The British Motor Show at Farnborough in August 2023 Source: Authors

being one of the first companies to produce a hydrogen fuel cell car – the Clarity, launched in 2008.

- **Korean Automakers**

In the spring of 2019 the Korean manufacturer Hyundai launched its hydrogen fuel cell SUV in the UK. An example of, this vehicle, the Hyundai Nexo was displayed at the British Motor show at Farnborough in 2023 by the breakdown service, the AA (Fig. 3.5). The AA is reported to be assessing the potential role for the vehicle in providing breakdown support to motorists in emissions-controlled parts of major cities (Salisbury 2022). Presumably it could be prohibitively expensive for the AA to operate older diesel fuelled vehicles in these cities.

In March 2024 it was reported that Hyundai and its compatriot Kia had joined forces to develop a hydrogen burning internal combustion engine (Newcomb, 2024).

3.3.2.2 New Entrant Case Study: Riversimple, UK

A British Disruptor

Riversimple is the result of more than 20 years' work by its founder Hugo Spowers (riversimple 2023a). Conscious of environmental sustainability he turned his back on conventional motorsport resolving to create a more sustainable form of personal mobility. In 2023 his philosophy is most clearly revealed by the Riversimple Rasa

Fig. 3.6 Professor Nuttall standing next to two Riversimple Rasa cars at the Hydrogen Gateway conference in Newport, Wales in June 2023. Source: Authors

hydrogen fuel cell car. Before we turn to the engineering choices that make possible the Rasa's impressive environmental credentials, it is worth commenting on the small car's funky look. While perhaps not to everyone's taste, the car has a certain style. With its scissor doors it makes a powerful style statement. This car is not afraid to say: *look at me!* (Fig. 3.6).

Engineering the Vision

The name of Riversimple's first fully-fledged car, the Rasa, comes from the Latin phrase "tabula rasa" literally meaning an erased tablet or perhaps more easily understood as a "clean slate" (Arstechnica 2016). The name symbolises that the intention to avoid any, and all, preconceived notions of what a future car should be, indeed Riversimple is so philosophical in its approach to innovation that the company refers to itself as a movement rather than a car manufacturer. The Rasa is, of course, a fully functioning car. Its principal engineering attributes are as follows:

- **Simple**

The name of the company includes the word – simple. Simplicity is part of the guiding philosophy at Riversimple. That simplicity can be seen in the design of the Rasa. If a car can be built that doesn't need a battery, then all the better. Through a combination of tried and tested fuel cell technology and innovative energy storage the Rasa has enough power to cruise on the highway and to accelerate to 60 mph in less than 10 s (riversimple 2023b). The main source of power to the wheels comes from Walmart 8.5 kW fork-lift fuel cells usually used to power fork-lift trucks in the warehouses of the American supermarket giant (Hydrogen Cars Now 2023). The Rasa has an extremely efficient regenerative braking system and that energy is not stored for later using a heavy battery, rather it is stored in a supercapacitor, from where it can rapidly be discharged providing acceleration when needed.

- **Lightweight and Aerodynamic**

The bodywork of internal combustion vehicles of the twentieth century needed to be strong before modern materials were available. Their petrol and diesel engines needed to withstand very high temperatures and be resistant to the erosion of hundreds of high-precision moving parts. It is not surprising therefore that cars were made of steel. As the twenty-first century got underway, in the UK at least, it seemed clear that the main way to lower the impact of personal mobility would be to reduce the weight of cars. New materials and construction techniques would make this possible with no detriment to passenger or pedestrian safety. Indeed, a prominent figure in forward thinking British automotive design, Gordon Murray has focussed on super-lightweight vehicle design as being key to the way ahead in the design of both performance sports cars and more day-to-day individual mobility. Gordon Murray Automotive declares: "*Lightweight design is a state of mind, and this approach helps deliver supercars with unmatched vehicle dynamics*" (Gordon Murray Automotive 2023). Like Gordon Murray, Hugo Spowers also had a background in British motorsport. Riversimple has turned its back on high-speed performance and focussed utterly on minimisation of environmental impact. Whether the focus is performance or efficiency, the core to the answer is the same. The cars of the twenty-first century must be lightweight.

In 1972 a petrol fuelled Austin Mini weighed roughly 620 kg (ultimateSPECS 2023). The completely redesigned "new" mini developed by BMW and launched in 2001 weighed only very slightly more than one Tonne (actually 1050 kg) for an early Cooper model (Evo 2023). Arguably, however, in recent decades trends in vehicle design have been going to other in the wrong direction – towards ever heavier cars for personal mobility. A modern Mini EV weighs more than 1.6 Tonnes.

As the years rolled by, concern for tailpipe emissions grew and battery electric vehicles found favour with policy makers. Tesla led the way, and indeed in so doing gave the world many wonderful innovations (for example in electric motor design) but light-weighting was not a Tesla gift to the world.

Indeed, Riversimple points out that the entire Rasa car (at a mere 655 kg) weighs roughly 1/3rd of a Tesla Model 3 Long-Range (at 1847 kg). In recent years concern has grown that the trend to battery electric vehicles (BEVs) is a move way from goodness. It simply cannot be right that in the early twenty-first century we move individuals in cars that weigh two Tonnes as many electric SUVs on the market from many manufacturers do today. That is simply wasteful of energy.

Why are BEVs so heavy? Charles Forsberg at MIT has a nice explanation. Forsberg points out that BEVs are so heavy because they carry their oxidising agent around with them. Both ICE powered vehicles and FCEVs extract oxygen from the air – they don't carry it around with them. Imagine a car having to lug around oxygen tanks and you get the idea behind Forsberg's pointed observation. That is basically what a BEV is doing – it may be "charged up" but the actual chemicals for both sides of the chemical reaction are carried on board the vehicle. It's also a reason that BEVs can present a fire risk that can be very hard to extinguish.

The Riversimple Rasa takes design in a totally different direction than another Tesla product – the Cybertruck. Riversimple focusses on minimising aerodynamic drag. It is bad enough having to accelerate that mass of a car without also having to force one's way through the air. By careful design the Rasa achieves a remarkably low drag coefficient (a measure of air resistance). Riversimple reports that the Rasa has a drag coefficient of 0.248 while in comparison the equivalent figure for a Porsche 911 is over 0.31.

- **Critical minerals**

Riversimple also points to growing concern for the critical minerals used in the manufacture of BEV battery packs. Because FCEVs are electric vehicles most FCEV developers imagine a combination of BEV and FEVs technologies going forward – see later discussion of Nikola trucks, for example. Riversimple, however, takes a different approach with the Rasa – it avoids the problems of batteries entirely by not having a battery. As noted above, the Rasa provides power for acceleration not through the use of batteries for energy storage (for example from regenerative braking) but rather through the use of supercapacitors. Supercapacitors are based on entirely separate scientific principles than batteries and they are manufactured in different ways from entirely different materials. Riversimple is able to avoid the changing geopolitics of energy by avoiding the use of scarce critical minerals from unstable parts of the world.

Practicality

The Riversimple Rasa provides zero-tail-pipe emissions personal mobility to those very many people that are not able to charge a battery electric vehicle conveniently and at low-cost. The Rasa can be refuelled in 3 minutes, as fast as a petrol or diesel car. This is an important consideration, as discussed elsewhere in this book [Chapter 4]. For commercial operators this could be particularly important consideration, it eliminates the issue of charging downtime seen with the BEV alternative, something that can be very harmful to established business models. It is a key reason that distribution warehouses turned to hydrogen FCEV forklift trucks establishing, and maturing, the core powertrain technology used by the Rasa.

Of course, the Rasa does not suit everyone's needs. It is a small two-seater car. It has limited storage space, but even more importantly it is not a good weight carrier. A super lightweight hi-tech car is probably not the best way to transport sacks of potatoes.

Funding

Riversimple has a personal approach to fundraising seeking support from enthusiasts and those that share the company's vision for more sustainable personal mobility (riversimple 2023c). Support from individual rich individuals through "family offices" has been boosted by a large number of small individual investments through the Seedrs crowdfunding platform. The other side to Riversimple's financial support has been a series of public grants and innovation support packages, including European Union support towards the construction of up to 20 prototype vehicles (Hydrogen Cars Now 2023). Thus far, as of August 2023, the company has remained independent of large industrial enterprises or financial institutions. The company continues to make its own future.

Business Model

Typically roughly 5.6 tonnes of CO_2 are emitted in the manufacture of a petrol or diesel car. Of that figure, about three quarters of which are released during production of the steel body (Auto Express 2023). Making a battery electric car typically releases even more CO_2 (roughly 8.8 Tonnes) of which more than 40% is associated with the manufacture of the battery (Auto Express 2023). These emissions are important as they equate to the emissions associated with driving a small conventional ICE petrol car more than 40,000 km. Any approach to sustainable personal mobility cannot simply be focussed on tailpipe emissions, it must be based on whole lifecycle emissions. Riversimple are clear they offer: *'mobility as a service, not cars as a product'*.

The business model is founded upon subscription, not purchase. For a single monthly fee everything is included (right up to fuel and insurance). In such a business model it is in everyone's interest to manufacture as few cars as possible and to make sure existing products can be refreshed and re-used for as long as possible.

As such Riversimple is not just a car company – it thinks of itself as a movement. The Riversimple movement aligns the interests of all stakeholders around virtues of freedom, fun, affordability and environmental responsibility. The fact that Riversimple is a hydrogen vehicle company is, in retrospect, almost incidental. That said hydrogen makes such a philosophy viable today in a way that seemingly no other technological option can match.

3.3.2.3 Closing Words on Hydrogen Cars

Hydrogen fuel cars have been developed from the earliest days of the twenty-first century by a range of major car manufacturers. So far none has achieved mainstream commercial success. Hydrogen FCEV technology has several key benefits over existing BEV alternatives, notably as concerns the needs of motorists unable to charge their vehicles at home using a cheap domestic electricity supply. That said, a hydrogen FCEV proposition requires consumer access to a network of filling stations for high purity hydrogen. While public and commercial BEV charging infrastructure continues to roll-out (despite the lengthy charging times) the roll-out of hydrogen supply for ordinary road users has not gone well. Shell has now pulled back in its hydrogen refuelling stations in the UK. In November 2022 it was reported that Shell would close its UK pilot hydrogen fuelling capability citing low levels of demand (Professional Driver Magazine 2022).

The delivery of the energy transition in transport and mobility is much more than just the delivery of the vehicles themselves. Supply chains and logistics coupled to a capable and commercially viable infrastructure are essential for success, and indeed for building consumer confidence. Despite the charging benefits of FCEV vehicles consumers might continue to avoid FCEV cars in the face of cheaper BEV alternatives supported by far more developed supportive infrastructure. Add to that Toyota's work on solid state batteries and hydrogen cars could indeed be in trouble. Perhaps Riversimple shows the way ahead with an emphasis on simplicity and above all lightweight design. At some level it is simply wrong that a car carrying just its 80 kg driver should weigh 2 Tonnes. Unless the new solid-state battery breakthrough is truly dramatic, one can expect that FCEV alternatives will always be lighter than BEV equivalents. With that said one is forced to admit that future cryogenic hydrogen tanks are a lightweighting concern. Importantly, such technology would be required in any future vehicle concepts relying on superconducting electric motors and hydrogen cryomagnetics. Such ideas will be discussed further in Chapter 5.

3.3.3   Two-Wheeled Vehicles

Thinking of two-wheeled vehicles immediately takes us to the emergent concept of micromobility. Micromobility is a term coined by Horace Dediu in 2017. It refers to individual short distance travel using minimally intrusive lightweight vehicles (Micromobility 2019). The vehicles are relatively low speed and include scooters and bicycles. The vehicles may be owned by their users, or be available for rent either commercially or as part of public provision. Typically, the vehicles are electrically powered or power-assisted. They are also frequently connected to electronic data networks. Dediu is an industry analyst with a particular background in mobile telecoms and associated systems. Indeed, he asserts that micromobility will be to mobility what microcomputing was for computing in the late twentieth century and up to the present day. A central issue is that the vehicle should be lightweight in complete contrast to the majority of electric cars on the road today. Indeed Dediu asserts that: the vehicle should weigh no more than five times the laden weight of the user (Micromobility 2019). Finally, the connection between micromobility and the notion of the smart city is a strong one and micromobility innovation is entirely compatible with ancient cities whose street layouts long predate the arrival of the automobile (Nuttall 2019).

Much of the micromobility has been enabled by the emergence of lightweight personal BEV technologies, such as eBikes and eScooters. In technical terms such technologies can be easily integrated into the networks of the smart city. What, if any, is the role for hydrogen powered micromobility?

One key issue that could come to favour the role of hydrogen in micromobility is fire safety. A key difference between BEV and FCEV approaches is that an FCEV requires oxygen from the atmosphere in order to function, whereas a battery contains all the components of the electrochemical reaction on board. This means that a battery fire is far harder to suppress. It will burn even in the absence of any external oxygen. Such concerns have seen eScooters banned from public transport across the world and there is growing concern around such devices being stored and charged in residential settings. Hydrogen powered alternatives would be refuelled at well equipped filling stations and would in the event of a fire be suppressed by more conventional means. With that said, however, hydrogen can leak and form combustible mixes with the air very easily. Furthermore, FCEV quality hydrogen must be pure and hence arguably free from odorants added for safety, although this is an area of active research around the world. The role of fire safety in micromobility technology choices is complex.

3.3.3.1   The Greening of Taipei's Scooter Culture

Per capita Taiwan is widely accepted to be the world's leading user of scooters for personal mobility (Wu 2019).

In an online magazine article David Woo has provided a good introduction to the historical importance of scooters to Taiwan (Wu 2019). The historical basis of the Taiwanese scooter economy has included the following elements:

- Personal ownership
- Petroleum fuels/internal combustion engines – noting very high fuel efficiency of low power scooters
- Original technology transfer from Japanese motorcycle companies
- Indigenous manufacturing

Looking to the future, it seems probable that Taiwan's affection for the scooter will continue, as Woo explains, but the exigencies of global climate change must surely lead to profound change. One can expect change to be driven, in part, by personal and policy concerns for greenhouse gas emissions and urban air quality. In the 1990s the Taiwanese government attempted to kick-start a shift to electric scooter use. In 2010 that initial push was described as a failure (Yang 2010).

The posited failure is attributed by Chi-Jen Yang to an excessive focus on subsidy and, in addition, we would suggest, an incomplete appreciation of other key factors, both technical and social. Twenty five years later we are seeing moves to electrified scooter adoption, but this entails a very wide range of consumer shifts and a move to wholly new business models. Looming is the high technology prospect of hydrogen-fuelled and fuel-cell powered electric scooters. Such a technology would bypass many of the perceived disadvantages of the battery electric alternative proposed since the 1990s. The prospect of a shift to hydrogen-based scooter mobility is not without technical challenges (network effects, safety concerns, economic viability, etc.). The shift to such technology, in a Taiwanese context, has already been envisaged for nearly 20 years, see Fig. 3.7, but may now finally be in prospect. Concerning the development of indigenous underpinning technology, we note with the progress of Asia Pacific Fuel Cell Technologies, located at Hsinchu Science Based Industrial Park, Chunan, Taiwan.

As noted elsewhere, orthodox hydrogen mobility thinking dictates that hydrogen transportation solutions will first be viable only in very heavy vehicles (e.g. trucks and buses etc.). The possibility of an emerging Taiwanese vision of ultra-lightweight hydrogen mobility is therefore radical. It will be interesting to see whether such

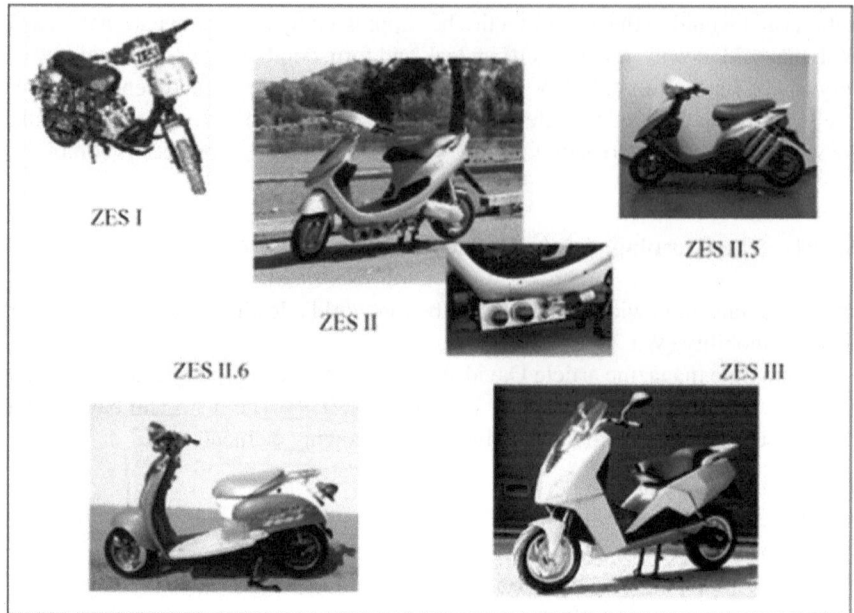

ZES I

ZES II.5

ZES II

ZES II.6

ZES III

Fig. 3.7 Early interest in Hydrogen fuel cell scooters in Taiwan, circa 2002. Source (Tso and Chang 2003) ccc Lisense, all rights reserved

ambitions succeed and what aspects (social, technical and economic) might motivate the emergence of a hydrogen-based micromobility approach. If it can happen anywhere then Taiwan is likely to be the place.

3.3.3.2   Hydrogen Motorcycles

Looking beyond the definition of micromobility, by allowing consideration of high-speed vehicles, one must pause to consider the potential for high-powered hydrogen fuelled motorcycles (BikeEXIF 2023). Both FCEV and ICE concepts are under consideration. The appeal of the ICE approach is, in part, a desire to preserve the noise and feel of today's high-powered technologies. It is felt, by some, that near silent BEV and FCEV technologies will not truly reflect the motorcycling experience for the rider, or indeed those nearby. Figure 3.8 (taken from Williams (2022)) shows the Kawasaki HySE concept motorcycle powered by a four-cylinder supercharged H2-ICE.

The parriers at the back of the bike are not to store the rider's belongings, that is where the hydrogen fuel is carried in this concept vehicle.

Fig. 3.8 Kawasaki HySE hydrogen fuelled ICE motorcycle concept. Source: (Williams 2022) Image copyright Kawasaki Motors, Ltd. All rights reserved

3.3.4 Buses

Around the world, and most especially in Europe, city governments have been keen to develop low carbon public transport. In the UK there are already several fleets of hydrogen powered buses in cities including: London, Aberdeen and Birmingham. The UK situation is summarised in Table 3.6 (taken from Zemo Partnership (2023)). By way of an admittedly imperfect comparison, the total number of Buses in Great Britain (i.e. excluding Northern Ireland) was 37,800 in 2020/21.

There are several reasons why hydrogen bus roll-out lags BEV technologies. First, and perhaps most importantly the issue is economics. Another major concern involves energy supply chains and refuel/recharge infrastructure. The availability of suitable vehicles remains a concern. For example, the major European bus manufacturing

Table 3.6 Low Carbon Buses in Service in the UK. Zero tailpipe emission technologies in bold. Data source: Zemo Partnership (2023). All rights reserved

Powertrain type	Number of buses in service
Hybrid	4572
Micro-Hybrid	2552
Battery Electric	**2330**
Biomethane	378
Hydrogen	**90**
Flywheel Hybrid	20
Plug-in Hybrid	6

company Salvador Caetano has developed variants of its CityGold low-floor single-deck bus fuelled by various alternative options, including: compressed natural gas and BEV. The company has produced a demonstrator Hydrogen FCEV variant for the UK market. This was trialled by Abellio in London in 2020.

Looking to the future, one might argue that while BEV technology might address well the easiest bus routes requiring a zero carbon solution, there could be many route patterns poorly suited to the BEV approach. With that said, there has been pioneering work in Milton Keynes, UK, concerning the wireless mid-route charging of BEV buses (Kontou and Miles 2015). Despite such endeavours, one can posit that, as the need grows to decarbonise the entire bus fleet, routes must be addressed that are poorly suited to the BEV solution. Routes that might be difficult for BEV approaches include those with long steep climbs or located in regions with cold winters among other more direct range-related factors. Concerning the economic challenge it is reported that the total cost of ownership for hydrogen FCEV buses will become cost competitive in the EU by 2035 (Deloitte China 2020).

While there are numerous examples around the world of successful deployments of BEV buses the potential for hydrogen is not lost. Indeed, in the summer of 2023 a public transport operator in the city of Rostock, Germany ordered 52 hydrogen FCEV buses as the first step in the complete replacement of its entire 170 vehicle fleet (Hydrogeninsight 2023). Enthusiasm for hydrogen buses is not just restricted to Europe. In August 2023 the Times of India reported the first hydrogen FCEV bus service entering into service in Leh a regional centre in the mountainous Ladakh region of northwest India (Times of India 2023).

3.3.5 Trucks

For an interesting overview of the potential for hydrogen in future low carbon European road haulage, the reader is recommended to consult a November 2021 report from Trafigura and H2 Energy (Trafigura and H2 Energy 2021).

In 2019 Patrick Molloy of the Rocky Mountain Institute in the US summarised well the case for hydrogen in heavy road road transport. Today the vast majority of trucks are fuelled with diesel, but to imagine a battery electric alternative brings some major issues including cost, loss of commercial carrying capacity and slow recharging times. The benefits of hydrogen are summarised in Molly's observations (Slanger 2019):

Diesel has an energy density of 45.5 megajoules per kilogram (MJ/kg), slightly lower than gasoline, which has an energy density of 45.8 MJ/kg. By contrast, hydrogen has an energy density of approximately 120 MJ/kg, almost three times more than diesel or gasoline. [...] 1 kg of hydrogen, used in a fuel cell to power an electric motor, contains approximately the same energy as a gallon of diesel.

Molloy further points to industrial suggestions that a hydrogen FCEV truck can be expected to obtain between 12 and 15 mpg equivalent, well above the US national average of approximately 6.4 mpg for a diesel fuelled truck.

Finally, Molloy observes that electric drivetrains, including FCEV technology, are also more efficient than diesel fuelled ICE alternatives. With an ICE, approximately 50% of the energy in the fuel is lost to heat; in contrast electric drivetrains, including those in FCEV vehicles, lose only about 10% to heat. PEM fuel cells in operation do typically produce some heat and this can be useful for heating the passenger compartment of an FCEV. Passenger cabin warmth can be a challenge in BEV design, especially for smaller vehicles such as cars. Generally, however, energy conversion efficiency from energy store to motion is a key advantage of hydrogen FCEV and BEV systems over diesel ICE. Many of these efficiency benefits of electrical energy conversion are independent of, and additional to, the low carbon advantages that would arise simply from fuel switching.

There is an interesting fact about replacing a diesel truck with a battery electric one. A study from the North American Council for Freight Efficiency (North American Council for Freight Efficiency 2023) suggests that, for the same weight of freight, this replacement would create a benefit of around 7,800 lbs (equivalent to 12% of the typical gross vehicle weight truck in U.S. roads). This would be possible by removing unneeded diesel components and fluids such as those related to engine and transmission, drive lines and axles, emission systems, as well as the diesel fuel itself.

In terms of carbon emission intensity, a report by IDTechEx (UK) (IDTechEx 2022) details the difficulty of Blue Hydrogen from natural gas reforming with CCS in seeking to match the low emissions targets that can be achieved by Green hydrogen. For instance, in the evaluation of different heavy-duty truck (HDT) powertrains according to type of technology, this study suggests that by 2030 carbon intensity (g/ CO_2/km) from FC-HDT with Blue Hydrogen would be around 4 times more than the ones using Green Hydrogen. When comparing emissions from BEV-HDT in EU by 2030, FC-HDT Green Hydrogen will still have the lowest emissions, this is because it is assumed that H_2 will be produced from 100% renewable electricity by 2030 while in EU the current assumption is for grid-average carbon emission intensity (Legatos 2023).

The availability of hydrogen fuelling stations is currently a barrier to expansion of this technology. As of January 2023 the number of hydrogen stations worldwide just exceeded 1,000. China is the country with the highest share (one third) followed by Japan. In Europe Germany is the only country with more than 100 installations, while in the UK there are 16 and 107 in USA (glpautogas 2023). California has the only state-wide hydrogen refuelling infrastructure in the U.S. Hydrogen FCEV sales slumped to less than 3,000 in 2022, relative to 280,000 for battery electric vehicles (Collins 2023). Hydrogen vehicle sales in California have been plagued by rising prices for hydrogen, where refuelling stations were (in 2022) charging greater than $20/kg (approx. $20/gallon of diesel equivalent) for hydrogen, (Klevstrand 2022).

The US Blueprint for Transportation Decarbonization (2023) shows no plan for expanded hydrogen utilization in light duty vehicles, which are assumed to be decarbonized by battery electric technology. The official US focus for hydrogen is on the long-haul heavy duty sector (Department of Energy 2023).

Honda is exploring options for vehicles (CR-V platform) which can switch between battery electric and fuel cell operation, to leverage BEV production to

Table 3.7 Summary analysis by McKinsey and Co. of four options for low-carbon trucking. The only identified significant weakness for hydrogen is the lack of a fuel supply infrastructure. Data Source: McKinsey & Company (2023)

	Bio/synfuel	Hydrogen internal combustion engines (H2-ICE)	Hydrogen (H2) fuel cell	Battery electric
Emissions				
CO_2 intensity	CO_2 intensity depends on source of biomass/carbon	Zero/minimal CO_2 if using Green/Blue H2	Zero/minimal CO_2 if using Green/Blue H2	CO_2 intesity depends on grid mix; zero CO_2 if using renewable power
Air quality	NOx[1] and particulate-matter emissions similar to diesel	No significant NOx emissions with SCR[2] after treatment	Zero emissions	Zero emissions
Total cost of ownership				
Efficiency (well-to-wheel)	~20%	~30% for renewable H2 production	~35% for renewable H2 production	75-85%+ depending on transmission and charging losses
Powertrain capital expenditure	Same as today's combustion engines	H2 engine with similar capex as diesel ICE, but H2 tank required	High capex for fuel cells and batteries, but more scalable than BEV[3]	High capex if large batteries required (medium for smaller/lighter segments)
Constraints (space/payload)	Same size and weight as today's combustion engines	Engine with same size as today, but H2 tank needed	More space needed than combustion engine for fuel cell and H2 tank	Higher weight than combustion engine; payload constraints subject to use case
Uptime/refueling	<15 minutes, depending on tank size	<15-30 minutes, depending on tank size	<15-30 minutes, depending on tank size	3+ hours, depending on ability for fast charging
Infrastructure costs	Can use existing infrastructure	H2 distribution and refueling infrastructure required	H2 distribution and refueling infrastructure required	Charging infrastructure and grid upgrades required

[1] Nitrogen oxides
[2] Selective catalytic reduction
[3] Battery electric vehicle

Variations across categories	High performance	Medium-high	Medium-low	Low Performance

accommodate the advanced features required for fuel cell versions (Collins 2022). A recent report from ITF (The International Transport Forum 2022) examines a total cost of ownership value proposition for H2 FCEV vs battery electric across all sectors. In EU markets, H2 FCEV are only cost competitive across a small fraction of the market sector. This causes risk that quantities of scale in hydrogen refuelling and OEM vehicle production may be too small to drive down costs (Table 3.7).

In May 2024 Volvo announced a plan to test hydrogen ICE trucks in 2026 (Hydrogeninsight 2024). The powertrain technology will employ high pressure direct injection. A small amount of hydrocarbon fuel will be needed to help start the truck engines (Hydrogeninsight 2024). In moving toward hydrogen ICE technology development Volvo is reported to be following a strategy already pioneered by German rival MAN (Transport News 2024).

3.3.5.1 New Entrant Case Study: Nikola, USA
A Troubled American Pioneer

Nikola is founded on the principle of providing environmentally responsible solutions for the north American road freight industry (Nikola 2023). As with Riversimple, discussed earlier in this chapter, a key part of the innovation is in the business model. Trucks are not sold nor owned. Haulage is provided as a service. Nikola reports that they have: *"developed an innovative, full-service lease program that covers the truck, the fuel, and the service"* (Nikola 2023).

The Vision

Nikola is taking on the hydrogen supply challenge through its HYLA business. HYLA reports that it takes a *"technology agnostic"* approach to hydrogen production (Nikola 2023), and as such it aligns to the philosophy of this book. HYLA is tackling all aspects for the hydrogen challenge from production, distribution (where it hopes to leverage existing infrastructures) and dispensing to FCEV trucks. Prior to launching the FCEV hydrogen fuelled trucking business, Nikola produced more than 200 Battery Electric Trucks on a similar business model to that proposed for the later FCEV business.

Turbulence

Despite being a young company, Nikola Corporation has had a turbulent history since being founded in Salt Lake City, Utah in 2014. By June 2020 the company's shares were trading at more than $60. But that apparent financial success was short-lived. The company was investigated by the activist short-seller Hindenburg Research and their report into alleged wrong doings crashed the share price in September 2020 (Hindenberg Research 2020). It did not recover in the subsequent three years and throughout the first half of 2023 the share price was always below $5 (in June 2024 the shares were rebased via a reverse stock split). The company parted ways with its founder after the Hindenburg research allegations. Since then, the company has seen a quick succession of leaders as it struggles to make forward progress. A second major setback came in August 2023, as the company confirmed a safety fault with its Tre BEV trucks. All 209 examples in use were recalled following the identification of a fault associated with battery cooling that had the potential to pose a fire risk (Inside EVs 2023). While this was undoubtedly further bad news for a troubled company, Nikola's long-awaited hydrogen fuel-cell trucks were reportedly unaffected by the fault and these started to be delivered to customers in late 2023.

In January 2024 a real world test was reported in which a Nikola FCEV truck hauled a 17.7 ton trailer for 400 miles over hilly terrain (Borrás 2024). The journey from the Port of Oakland to the Port of Los Angeles in Long Beach California involved driving through the Tejon Pass at 1268 m above sea level. The driver, William Hall, who is also founder of the company involved (Coyote Container), described the experience: *"The truck is a dream to drive and I arrived at Pier C with about 140 miles of remaining range, [...] I drove conservatively and did the steep climbs at 40 mph, much as I would in a diesel with that load profile, though I could have easily gone 55 (mph)."*

In the case of Nikola one can take the view that while the central premise of hydrogen FCEV heavy transport makes sense, especially in the North American market, Nikola as a company has arguably not done well in building trust with investors or potential customers. Whether Nikola will emerge as the leader of a new hydrogen trucking revolution remains far from certain. A once much celebrated equity stake by US automotive giant GM in 2020 was liquidated once the Hindenburg Research report was published, but GM was, in January 2021, according to reports still partnering with Nikola on FCEV technology development (The Verge 2021). The main focus of GM's interest turned however to Nikola's competitor Navistar, owner of the "International" brand for trucks and diesel engines, based in Lisle, Illinois. Perhaps Navistar and GM as examples of major American corporations will be the ones to lead North American heavy haulage into a low carbon future based on hydrogen. As the Danish physicist, Niels Bohr, and others, are reported to have said: *'prediction is difficult, especially about the future'*.

3.3.6 Specialist Vehicles

So far, we have focussed on vehicles designed to be used on public roads, but off-road vehicles can be a fertile sector for hydrogen innovation. Often the vehicles are operated by highly technically competent firms comfortable with the engineering issues of moving to a new powertrain technology. Such operators are also well-placed to control the supply chains and maintenance requirements of new vehicles. One prominent example of such innovation was led by the mining giant Anglo American which is exploring ways to switch the power train from its large mining vehicles to hydrogen FCEV technology. Anglo American has an added incentive to see the development of hydrogen mobility as the company explains:

> As one of the world's leading producers of platinum group metals (PGMs), we were early supporters of the global potential for a hydrogen economy, recognising its role in enabling a shift to more sustainable energy and cleaner transport. PGMs are used as catalysts in Fuel Cell Electric Vehicles, which offer many benefits to the transport sector, including refuelling times comparable to internal combustion engine vehicles, long ranges, and space and weight efficiency. Our integrated approach includes investing in new technologies, supporting entrepreneurial projects, and advocating for policy frameworks that enable a supportive long-term investment environment for hydrogen to deliver on its potential. Source: (AngloAmerican 2023)

In the UK construction equipment manufacturer JCB has replaced the diesel powertrain in a truck with a new replacement liquid hydrogen fuelled FCEV drive-train (Wu 2019). JCB's interests in hydrogen powertrain manufacture extends beyond its UK operations. In September 2023 it was reported that JCB seeks to build hydrogen ICE power units in India (Krishnan 2023). The company operates six manufacturing facilities across India from which it exports 45% of its products outside India.

As we have seen, JCB is not alone in seeing potential for hydrogen for large off-road specialist vehicles. The steps taken by the mining giant Anglo American point to a future where large manufacturers and operators of specialist vehicles might make a switch to a zero emission alternative at their own initiative. Such innovations can be particularly interesting as, unlike with publicly funded R&D or those innovations prompted directly by public policy initiative, these technological developments only become visible once they are complete, or nearly so.

3.3.7 Road Vehicles Conclusions

Many European, or British, observers might say that there has been a battle between BEV and FCEV technologies and BEV has won. We suggest, however, that it is too early to say that for FCEV the battle is lost. For example one of our panellists, Michaela Kendall advised the authors that *"despite growth in electric battery vehicles, a recent survey found that 79% of automotive directors believe that fuel cell vehicles will be the real breakthrough for electric mobility."* – The relevant survey (originally published by KPMG) has also been widely cited by other experts including (H2FCSUPERGEN 2023).

We have stressed in this section the importance of various factors that have the potential to favour hydrogen technologies in the future. These include:

- Rapid refuelling
- The virtue inherent in lightweighting personal mobility.
- The differing safety attributes of BEV and FCEV approaches.
- The shifting relative attributes of BEV and FCEV approaches as we attempt to decarbonise ever more difficult challenges.
- The shifting economics, driven largely by supply chains and infrastructures.
- The attitudes and strategies of large industrial incumbents.

As we consider the attributes above it is important to reflect on the fact that BEV vehicles are far from being a zero environmental impact proposition (Oğuz 2023). Looking ahead it becomes possible to imagine lighter and simpler FCEV vehicles for which whole life-cycle emissions could be reduced significantly compared to the conventional, hybrid and BEV options listed (Oğuz 2023). Of course, as noted earlier, innovation is occurring in many areas simultaneously. While hydrogen vehicles might innovate themselves to the front of the queue, it is also possible that BEV technologies will see breakthrough innovation (such as the solid-state batteries under development by Toyota) that could effectively put an end to hydrogen as a ground transportation option.

3.3.8 Other Vehicles

3.3.8.1 Trains

There is potential for hydrogen powered trains to replace the existing diesel stock to help achieve net zero. Traditionally train lines were electrified to achieve this, however it may be cheaper running hydrogen powered rolling stock rather than further investing in new infrastructure for electrified train lines. This is especially important for those track routes where relatively little traffic is expected. Indeed, hydrogen could well be the best option for trains in certain locations that are difficult to electrify (Fuel Cells and Hydrogen 2 Joint Undertaking 2019).

Alstom has shown the efficacy of hydrogen-powered trains with demonstrations in Germany and France, leading to initial commercial deployments (Buckley 2022). Battery electric options generally cannot provide the on-board storage needed to travel between stops in commuter and other rail services. Therefore, until recently electrification has required the installation of third-rails or catenary overhead lines to provide continuous power along the route. These traditional electrification options are expensive and unattractive to stakeholders relative to H2 refuelling with onboard storage. After an extensive comparison study, San Bernadino County Transit Authority SBCTA) chose hydrogen fuel cell over electrified train options for the first Hydrogen Zero Emission Train in North America (Buckley 2022). As a consequence of such logics, various countries around the world are testing, or even operating, hydrogen powered trains. Figure 3.9 (taken from Hackaday (2023)) shows an example from China which makes use of hydrogen energy and supercapacitor electricity storage, in a manner similar to the Riversimple Rasa car, discussed earlier. Despite its modest top speed, it was in fact at that time the fastest hydrogen powered train in the world (Hackaday 2023).

In September 2023 it was reported that Irish Rail in collaboration with Latvian partners, DIGAS, is planning to convert an existing diesel locomotive such that it operates by combusting hydrogen in place of diesel fuel (Railtech 2023). The aim is

Fig. 3.9 CRRC 160 km/hour zero emission hydrogen train in February 2023. Source: (Hackaday 2023) Image copyright CRRC. All rights reserved

to convert the existing ICE for the new clean fuel as a proof of concept. In 2023 the work is still at the design stage with engineering testing planned for 2024 and 2025.

While China pushes ahead with hydrogen powered FCEV rail ambitions progress is not without its problems in Europe. In August 2023 it was announced that in Lower Saxony, Germany the regional government would end its use of 14 hydrogen trains that it started running in August 2022. It is reported that BEV alternative service delivery is simply available at lower cost (Quartz 2023). It remains to be seen the extent that BEV technology can address rail transport demand in more challenging geographical and market contexts. Heavy rail freight in north America and Australia is a very different proposition that light commuter rail in Germany.

3.3.8.2 Shipping

Shipping is a complex area involving some fast-moving technological developments. One point of possible confusion is that hydrogen and shipping come together for two reasons. The first reason is analogous to the other transport technologies discussed above. Can hydrogen act as a low carbon fuel for today's merchant fleet and hence break the link to heavy fuel oil usage in marine propulsion. Civil marine propulsion is one of the most difficult sectors to decarbonise and hydrogen (or a related synfuel) could well have a beneficial role to play. For example, in November 2023 it was reported that three companies Yara Clean Ammonia, NorthSea Container Line and Yara International are collaborating to develop the world's first container ship that will use pure ammonia as fuel (Central 2023). The vessel, to be known as the Yara Eyde will operate between Oslo, Brevik in Norway and Hamburg and Bremerhaven in Germany from 2026.

The second point of commonality is rather different. It concerns the future prospects for a global trade in hydrogen and related low carbon synthetic fuels, including, for example ammonia as planned as a fuel for the Yara Eyde. The issue of low carbon fuels as future shipping cargoes is a distinct issue, and an area of active research for members of our group. In collaboration with our colleague Marc Cochrane, we expect some relevant research outputs to be published soon.

In a scenario where hydrogen is shipped globally, the cargo can be used as a zero-carbon emissions fuel. In the case of cryogenic liquid hydrogen shipping, for example, this is the possibility of using boil-off from the cargo as propulsion fuel rather than expending energy to reliquefy it. Hydrogen liquefaction is a relatively difficult technical proposition, as discussed in chapter 5. Such an approach to boil-off utilisation would mimic aspects of how LNG is shipped globally. Fundamentally we expect international hydrogen trade to come before hydrogen shipping. Once the trade is underway at scale and depending on the nature of the cargo (for example a carrier molecule might be used) then there is potential to link to separate developments seeking to utilise hydrogen, or a related synfuel, as the primary fuel for shipping propulsion, as mentioned earlier.

3.3.8.3 Aviation

Of all the transport and mobility issues we suggest that aviation could be the key to hydrogen energy commercialisation. The need to decarbonise aviation is manifest, but the challenge is severe. Great efforts are underway to achieve a net zero result by building on today's sustainable aviation fuels (essentially hydrocarbons derived from renewable biomass). Irrespective of the merits of SAF operators and customers are likely to be attracted to providers who can validly claim that their aircraft emit no greenhouse gases from the engines. It is important that the manufacture of SAF itself requires large amounts of hydrogen – see the discussion of E-fuels in Chapter 2.

The weight of battery packs currently renders such an approach non-viable for any large passenger or cargo aircraft. A battery powered plane might fly, but can it also carry a cargo?

As fuel cells and electric motors are lighter than HICE it is likely that this will be the preferred technology in the aeronautical industry. Prior to its June 2024 collapse, Universal Hydrogen proposed aeroplanes based on existing models but with a hydrogen fuel cell powertrain available for new aircraft. That company planned to sell conversion kits for regional aircraft like the De Havilland Canada Dash-8 (Universal Hydrogen 2023).

Another pioneer in this space is Zeroavia (CNBC 2023). That company is developing medium-sized hydrogen powered planes that offer commuter and transport options that are not possible using battery electric technology (ZeroAvia 2023). Airbus has an extensive Zero-e program to examine turbofan, turboprop, and blended-wing body for future hydrogen-powered design (Airbus 2021). The potential to use low temperature (20 K) liquid hydrogen to enable high efficiency superconductivity engine designs is under examination, which could be a game changer for powered flight see chapter 5, Box 5.4. That said metallurgical considerations must be mentioned. First any leakage of hydrogen could cause hydrogen molecules to enter into the structural metals of the aircraft – risking potential embrittlement. Secondly the use of cryogenics on an aircraft also runs the risk of metallic fracture of ultra-cold components. Aviation structures need to be both flexible and strong. Any threat to metallic flexibility would be a serious design concern. Nevertheless, Airbus sees a future for hydrogen hybrid turboprop powered aircraft carrying up to 100 passengers over distances up to 1,000 nautical miles or more. For longer journeys up to 2,000 nautical miles or more, 200 passengers could be carried in a relatively conventional aircraft body concept or a more radical blended-wing body design (Airbus 2021). The longer-range options will be powered by a pair of hydrogen hybrid turbofan engines. Issues of hydrogen in aviation will be explored further in Chapter 5. Blended wing bodies have been under development for far longer than hydrogen has been under active consideration as an aviation fuel. An example of a blended wing body aircraft concept is shown in Fig. 3.10.

There are reports of pioneering entrepreneurs preparing to introduce hydrogen fuelled aviation to the travelling public (The Guardian 2023). Those seeking to roll-out commercial services will be conscious of hydrogen supply chains (and the need to ensure high levels of quality and reliability). Another consideration is refuelling

Fig. 3.10 Rendering of a US Air Force blended wing body aircraft concept to be progressed by JetZero pioneers in blended wing body design (U.S. Air Force 2023)

times. The rise of low-cost airlines to the point that today they dominate short and medium haul aviation reminds us that a plane on the ground is not making money. In principle hydrogen has the potential to preserve such business models.

3.4 Conclusions

In this chapter we have considered the context of hydrogen for transport applications and various companies pursuing innovative ambitions. In Chapter 4 we shall consider hydrogen developments around the world. In Chapter 5 we shall consider further important emerging hydrogen ideas, some relating to transport and mobility. In Chapter 6 we shall offer final comments and conclusions. Despite that role for Chapter 6, it is worth pausing here to reflect on some of the concepts explored above.

As regards the use of new fuels in the global shipping of freight. We heard in our workshops that the commodity supply industry is actively looking at both green ammonia and green methanol as potential future low-carbon shipping fuels. One issue of concern is the size of the fuel tanks needed. For example, the study has heard that ammonia fuels need roughly double the fuel tank size when compared to conventional heavy fuel oil. Electric ship solutions are also interesting, but it is believed, that the greatest potential there will be for very small cargo ships (such as river barges) and passenger ferries in the area of mobility. One potential idea, for the largest ocean-going cargo vessels would be on-board fossil fuel to syn-fuel processing with CCS. One example would be to use liquefied natural gas (LNG) fuel, on board reformation perhaps via methods, such as microwave plasma processing, that might yield a solid carbon waste-form for in-port discharge.

The potential for innovation in global shipping is large and the study has heard that commodity sector leading player, Trafigura is pushing for a global \$300 per Tonne CO_2 carbon price for shipping. The company says that at such a carbon price dramatic decarbonisation can result. We would add that for progress to be made such a price must be paid by all those involved in what is today a global competitive market. The challenge will be to achieve such a global price while preserving the wider economic benefits of competition.

We observe that hydrogen fills a potential gap appearing between on the one hand electric vehicles and on the other synthetic cleaner hydrocarbon fuels/biofuels. We note that there is at each step a 50% drop in efficiency first from going from Battery Electric Vehicle (BEV) to hydrogen-based FCEV to Biofuel options. The biofuel options in internal combustion are the options closest to today's diesel-based road haulage system. The shift to FCEV and BEV options is a beneficial shift in terms of emissions generation, but also in terms of system efficiency. The BEV alternative is attractive, especially for smaller vehicles such as bikes and cars, it remains to be seen if the efficiency advantages will allow BEV technology to displace the anticipated role of FCEV in low-carbon heavy transport. Energy, or even economic, efficiency alone will not be the only driver, however, another key driver of change will be the changing nature of embedded infrastructure. It is in the area of infrastructure that private sector leadership runs the risk of being constrained by the actions of government. That said, one should not seek to avoid government's role; rather there is a need for industry and government to work together to establish priorities and, above all, to move quickly as 2050 is, infrastructure terms, an imminent date.

To close, it remains important to concede, that in transport and mobility at least, hydrogen remains a contested proposition. Hydrogen's main technological rival is battery energy storage where there is also much innovation underway. For current applications, at our current level of decarbonisation, generally hydrogen is an unattractive choice. Hydrogen advocates must continue to explore low-cost options for production and distribution. In this regard the emergence of low-cost geological (White) hydrogen could be especially important especially in some territories. Otherwise, the focus must be to continue to reduce costs for both green and blue hydrogen, while continuing to explore more advanced technological options (Red hydrogen etc.).

Geography is crucial and we shall consider that aspect further in the next chapter. Innovation is the other key consideration, and we shall return to that theme in Chapter 5.

References

@rechargenews (2022) Hydrogen car sales almost doubled last year—after drivers were offered 50–65% discounts. 2022-02-14 [cited 2023 30 September]. https://www.rechargenews.com/energy-transition/hydrogen-car-sales-almost-doubled-last-year-after-drivers-were-offered-50-65-discounts/2-1-1168221

Airbus (2021) ZEROe: towards the world's first hydrogen-powered commercial aircraft. 2021-06-24 [cited 2023 20 August]. https://www.airbus.com/en/innovation/low-carbon-aviation/hydrogen/zeroe

Alternative Fuels Data Center (2023) How do fuel cell electric vehicles work using hydrogen? [cited 2023 23 October]. https://afdc.energy.gov/vehicles/how-do-fuel-cell-electric-cars-work

AngloAmerican (2023) Hydrogen: here for the long haul. [cited 2023 22 August]. https://www.angloamerican.com/our-stories/innovation-and-technology/hydrogen-here-for-the-long-haul

Arstechnica (2016) Riversimple Rasa review: is this hydrogen car the future—or just a gimmick? 2016-04-20 [cited 2023 19 August]. https://arstechnica.com/cars/2016/04/riversimple-rasa-hydrogen-car-review/

Auto Express (2023) Car pollution facts: from production to disposal, what impact do our cars have on the planet? [cited 2023 19 August]. https://www.autoexpress.co.uk/sustainability/358628/car-pollution-production-disposal-what-impact-do-our-cars-have-planet

Bauer C et al (2022) On the climate impacts of blue hydrogen production. Sustainable Energy Fuels 6(1):66–75

BBC (2015) Volkswagen: the scandal explained. [cited 2023 30 September]. https://www.bbc.co.uk/news/business-34324772

BikeEXIF (2023) Hydrogen powered: could hydrogen power your next bike? 2023-02-02 [cited 2023 20 August]. https://www.bikeexif.com/?p=70331

BMW (2023a) Step inside the new BMW iX5 Hydrogen. [cited 2023 22 August]. https://discover.bmw.co.uk/article/step-inside-the-new-bmw-ix5-hydrogen

BMW (2023b) BMW iX5 Hydrogen pilot fleet visits the UK. [cited 2023 30 September]. https://www.press.bmwgroup.com/united-kingdom/article/detail/T0409829EN_GB/bmw-ix5-hydrogen-pilot-fleet-visits-the-uk?language=en_GB

Borrás J (2024) Nikola hydrogen FCEV semi completes 400 mile trip—is it enough? 2024-01-29 [cited 2024 7 June]. https://electrek.co/2024/01/29/nikola-hydrogen-fcev-semi-completes-400-mile-trip-is-it-enough/

Buckley J (2022) The world's first hydrogen-powered passenger trains are here. 2022-08-24 [cited 2023 20 August]. https://www.cnn.com/travel/article/coradia-ilint-hydrogen-trains/index.html

California Energy Commission (2019) Joint Agency Staff Report on Assembly Bill 8: 2019 Annual Assessment of Time and Cost Needed to Attain 100 Hydrogen Refueling Stations in California

Chemistry World (2023) The seabirds saved by synthetic chemistry. https://www.chemistryworld.com/opinion/the-seabirds-saved-by-synthetic-chemistry/4014400.article

CNBC (2023) Why Airbus and others are betting on hydrogen-powered planes instead of electric planes. 2023-05-11 [cited 2023 20 August]. https://www.cnbc.com/2023/05/11/why-airbus-and-others-are-betting-on-hydrogen-powered-planes.html

Collins L (2022) 'Exploring potential' | Honda announces new FCEV for US market that can be switched between hydrogen and battery operation. 2022-11-30 [cited 2023 17 November]. https://www.hydrogeninsight.com/transport/exploring-potential-honda-announces-new-fcev-for-us-market-that-can-be-switched-between-hydrogen-and-battery-operation/2-1-1364079

Collins L (2023) Hydrogen fuel-cell car sales slumped in California in 2022 as battery EVs boomed. 2023-01-11 [cited 2023 17 November]. https://www.hydrogeninsight.com/transport/hydrogen-fuel-cell-car-sales-slumped-in-california-in-2022-as-battery-evs-boomed/2-1-1386678

Crouch D (2023) Toyota reveals British-built, hydrogen-fuelled Hilux Prototype pick-up. 2023-09-04 [cited 2023 6 November]. https://media.toyota.co.uk/toyota-reveals-british-built-hydrogen-fuelled-hilux-prototype-pick-up/

Davies R (2023) Toyota claims battery breakthrough in potential boost for electric cars. 2023-07-04 [cited 2023 22 August]. https://www.theguardian.com/business/2023/jul/04/toyota-claims-battery-breakthrough-electric-cars

Deloitte China (2020) Hydrogen and fuel cell solutions for transportation

Department of Energy (2023) The U.S. National Blueprint for Transportation Decarbonization

Derwent R et al (2006) Global environmental impacts of the hydrogen economy. International Journal of Nuclear Hydrogen Production and Applications 1(1):57–67

Dronsfield A (2023) Who really discovered the Haber process? [cited 2023 12 October]. https://edu.rsc.org/feature/who-really-discovered-the-haber-process/2020277.article

E4tech (2016) Hydrogen and fuel cells: opportunities for growth—a roadmap for the UK

Eco News (2024) New hydrogen engine with unprecedented injection: 2 liters and 40% thermal efficiency. 2024-05-03. https://www.ecoticias.com/en/new-hydrogen-engine-injection/1596/

Equinor (2020) H2H Saltend

Eurostat (2023) Statistics explained. [cited 2023 23 October]. https://ec.europa.eu/eurostat/statistics-explained/index.php?title=Energy_consumption_in_households

Evo (2023) Mini Cooper (R50) and Cooper S (R53)—history, prices and specs. [cited 2023 19 August]. https://www.evo.co.uk/mini/cooper/19635/mini-cooper-r50-and-cooper-s-r53-history-prices-and-specs

Fuel Cells and Hydrogen 2 Joint Undertaking (2019) Hydrogen roadmap Europe

glpautogas (2023) Hydrogen stations in the world. Map and List. [cited 2023 17 November]. https://www.glpautogas.info/en/hydrogen-stations.html

Gordon Murray Automotive (2023) Ethos/lightweighting. [cited 2023 19 August]. https://www.gordonmurrayautomotive.com/

H2 View (2020) Preview: shell exclusive on its latest mega-project, the NortH2 green hydrogen plan. https://www.h2-view.com/story/preview-shell-exclusive-on-its-latest-mega-project-the-north2-green-hydrogen-plan/

H21 (2019) H21 Leeds city gate

H2FCSUPERGEN (2023) Opportunities for hydrogen and fuel cell technologies to contribute to clean growth in the UK. [cited 2023 22 August]. https://www.h2fcsupergen.com/opportunities-for-hydrogen-fuel-cell-clean-growth-uk/

Hackaday (2023) China's new 100 MPH train runs on hydrogen and supercaps. 2023-02-02 [cited 2023 20 August]. https://hackaday.com/2023/02/02/chinas-new-100-mph-train-runs-on-hydrogen-and-supercaps/

Hindenberg Research (2020) Nikola: how to parlay an ocean of lies into a partnership with the largest auto OEM in America. [cited 2023 20 August]. https://hindenburgresearch.com/nikola/

Hydrogen and Fuel Cell Strategy Council (2019) The strategic road map for hydrogen and fuel cells

Hydrogen Central (2023) Yara—the world's first ammonia powered container ship. 2023-11-03 [cited 2023 5 November]. https://hydrogen-central.com/yara-the-worlds-first-ammonia-powered-container-ship/

Hydrogen Cars Now (2023) Riversimple Rasa review. [cited 2023 19 August]. https://www.hydrogencarsnow.com/index.php/riversimple-rasa/

Hydrogeninsight (2023) German public transport company orders 52 hydrogen buses for €30m after rejecting battery-electric models. 2023-04-21 [cited 2023 22 August]. https://www.hydrogeninsight.com/transport/german-public-transport-company-orders-52-hydrogen-buses-for-30m-after-rejecting-battery-electric-models/2-1-1438416

Hydrogeninsight (2024) Volvo to roll out first trucks with hydrogen internal combustion engines in 2026. https://www.hydrogeninsight.com/transport/volvo-to-roll-out-first-trucks-with-hydrogen-internal-combustion-engines-in-2026/2-1-1648604

HyNet (2018) HyNet North West: from vision to reality

IDTechEx (2022) Fuel cell electric vehicles 2022–2042

IEA (2023) Methane abatement options—methane tracker 2020—analysis. https://www.iea.org/reports/methane-tracker-2020/methane-abatement-options

Inside EVs (2023) Nikola recalls Tre BEV truck, temporarily stops sales over battery safety issue. [cited 2023 20 August]. https://insideevs.com/news/681667/nikola-tre-bev-recall-stop-sale/

International Energy Agency (2015) Technology roadmap: hydrogen and fuel cells

International Energy Agency (2019) The future of hydrogen

Kendall M, Kendall K, Lound APB (2023) HYSTORY: the story of hydrogen - Adelan SOFC. @AdelanLtd

Klevstrand A (2022) Fresh blow for hydrogen vehicles as average pump prices in California rise by a third to all-time high. 2022-11-10 [cited 2023 17 November]. https://www.hydrogeninsight.com/transport/exclusive-fresh-blow-for-hydrogen-vehicles-as-average-pump-prices-in-california-rise-by-a-third-to-all-time-high/2-1-1351675

Koch T (2022) Hydrogen reality check #1: hydrogen is not a significant warming risk. 2022-05-09 [cited 2023 1 December]. https://rmi.org/hydrogen-reality-check-1-hydrogen-is-not-a-significant-warming-risk/

Kontou A, Miles J (2015) Electric buses: lessons to be learnt from the Milton Keynes Demonstration Project. Procedia Eng 118:1137–1144

Krishnan J (2023) JCB's hydrogen powered vehicles become a reality, awaiting mass production. 2023-09-10 [cited 2023 6 November]. https://www.thehindubusinessline.com/news/jcbs-hydrogen-powered-vehicles-become-a-reality-awaiting-mass-production/article67291710.ece

Legatos J (2023) Hydrogen truck demo lets Alberta carriers do real-world test drives. 2023-08-16 [cited 2023 22 August]. https://electricautonomy.ca/2023/08/16/hydrogen-truck-demo-alberta-carriers/

McKinsey & Company (2020) Decarbonization challenge for steel

McKinsey & Company (2023) How hydrogen combustion engines can contribute to zero emissions. https://www.mckinsey.com/industries/automotive-and-assembly/our-insights/how-hydrogen-combustion-engines-can-contribute-to-zero-emissions

Micromobility (2019) The micromobility definition. [cited 2023 19 August]. https://micromobility.io/news/the-micromobility-definition

Newcomb T (2024) A New Hydrogen Combustion Engine Is Legitimately Heating Up, Popular Mechanics. [cited 2024 29 May]. https://www.popularmechanics.com/science/green-tech/a60915727/hydrogen-combustion-engine-innovation/ accessed 7 October 2024.

Nikola (2023) Energy—pushing the boundaries of possibility. [cited 2023 20 August]. https://www.nikolamotor.com/energy/

North American Council for Freight Efficiency (2023) Electric trucks. [cited 2023 17 November]. https://nacfe.org/research/electric-trucks/#viable-heavy-duty-electric-tractors

Nuttall WJ (2019) Historic European cities and the smart future. In: Energy and mobility in smart cities, pp 19–29

Nuttall WJ, Bakenne AT (2020) Fossil fuel hydrogen. Springer, Bern

Office of Energy Efficiency & Renewable Energy (2021) Hydrogen shot summit. [cited 2022 20 June]. https://www.energy.gov/eere/fuelcells/hydrogen-shot

Oğuz S (2023) Life cycle emissions: EVs vs. combustion engine vehicles. 2023-06-23 [cited 2023 20 August]. https://www.visualcapitalist.com/life-cycle-emissions-evs-vs-combustion-engine-vehicles/

Pieterse et al (2013) Reassessing the variability in atmospheric H2 using the two-way nested TM5 model. Journal of Geophysical Research: Atmospheres 118(9):3764–3780

Professional Driver Magazine (2022) Shell closes UK hydrogen filling stations due to lack of demand. https://www.prodrivermags.com/ [cited 2023 20 August]. https://www.prodrivermags.com/news/shell-closes-uk-hydrogen-filling-stations-due-to-lack-of-demand/

Quartz (2023) The dream of the first hydrogen rail network has died a quick death. 2023-08-07 [cited 2023 20 August]. https://qz.com/the-dream-of-the-first-hydrogen-rail-network-has-died-a-1850712386

Railtech (2023) Irish Rail to trial Europe's first locomotive to combust hydrogen instead of diesel. [cited 2023 6 November]. https://www.railtech.com/innovation/2023/09/19/irish-rail-and-latvias-digas-to-trial-europes-first-retrofitted-hydrogen-freight-locomotive/

Reuters (2023) Honda to start producing new hydrogen fuel cell system co-developed with GM. 2023-02-02. https://www.reuters.com/business/autos-transportation/honda-start-producing-new-hydrogen-fuel-cell-system-co-developed-with-gm-2023-02-02/

riversimple (2023a) History of vehicle development. [cited 2023 16 August]. https://www.riversimple.com/history-of-vehicle-development/

riversimple (2023b) Innovation. [cited 2023 19 August]. https://www.riversimple.com/innovation/

riversimple (2023c) investors. [cited 2023 14 August]. https://www.riversimple.com/investors/

Robiniusa M et al (2018) Comparative analysis of infrastructures: hydrogen fueling and electric charging of vehicles. Institut für Elektrochemische Verfahrenstechnik

Salisbury M (2022) AA unveils hydrogen patrol vehicle at British Motor Show 2022. 2022-08-22. https://fleetpoint.org/hydrogen-vehicles/aa-unveils-hydrogen-patrol-vehicle-at-british-motor-show-2022/

Sand M et al (2023) A multi-model assessment of the global warming potential of hydrogen. Communications Earth & Environment 4(1):203

Slanger D (2019) Run on less with hydrogen fuel cells—RMI. 2019-10-02. https://rmi.org/run-on-less-with-hydrogen-fuel-cells/

Statistica (2023a) Global demand for agricultural fertilizer by nutrient from 2011/2012 to 2022/2023. [cited 2023 30 August]. https://www.statista.com/statistics/438930/fertilizer-demand-globally-by-nutrient/

Statistica (2023b) Ammonia production worldwide from 2010 to 2022. [cited 2023 30 September]

Steel Times International (2024) Nippon Steel cuts carbon emissions by a third through hydrogen trial. [cited 2024 5 June]. https://www.steeltimesint.com/news/nippon-steel-cuts-carbon-emissions-by-a-third-through-hydrogen-trial

T3 (2023) The BMW iX5 Hydrogen proves that batteries aren't the only answer

The Federal Government (2020) The national hydrogen strategy

The Fuel Cell and Hydrogen Energy Association, Road Map to a US Hydrogen Economy

The Guardian (2023) Green energy tycoon to launch UK's first electric airline. 2023-07-16 [cited 2023 20 August]. https://www.theguardian.com/business/2023/jul/17/green-energy-tycoon-to-launch-uk-first-electric-airline

The International Transport Forum (2022) Decarbonising Europe's trucks. OECD Publishing, Paris

The Nobel Prize (2023) The Nobel Prize in chemistry 1918. https://www.nobelprize.org/prizes/chemistry/1918/haber/biographical/

Times of India (2023) India's first hydrogen fuel cell bus service to hit the roads in Leh

Trafigura and H2 Energy (2021) Decarbonising heavy-duty trucking and accelerating the European hydrogen economy

Transport News (2024) Volvo reveals hydrogen ICE truck plans. 2024-05-24 [cited 2024 7 June]. https://www.transportnews.co.uk/2024/05/24/volvo-reveals-hydrogen-ice-truck-plans/

Tso C, Chang S-Y (2003) A viable niche market—fuel cell scooters in Taiwan. Int J Hydrogen Energy 28(7):757–762

U.S. Air Force (2023) DAF selects JetZero to develop blended wing body aircraft prototype. [cited 2023 7 November]. https://www.af.mil/News/Article-Display/Article/3494520/daf-selects-jetzero-to-develop-blended-wing-body-aircraft-prototype/

U.S. Energy Information Administration (2021) Electric power sector CO2 emissions drop as generation mix shifts from coal to natural gas. [cited 2023 12 October]. https://www.eia.gov/todayinenergy/detail.php?id=48296

ultimateSPECS (2023) 1972 Austin Mini 1000 specs, dimensions. [cited 2023 19 August]. https://www.ultimatespecs.com/car-specs/Austin/376/1972-Austin-Mini-1000.html

Universal Hydrogen (2023) Product. [cited 2022 29 September]. https://hydrogen.aero/product/

US Energy Information Administration (2022) Country analysis executive summary: Australia. US Energy Information Administration

US Energy Information Administration (2023) Hydrocracking is an important source of diesel and jet fuel. https://www.eia.gov/todayinenergy/detail.php?id=9650

The Verge (2021) General Motors, after flubbing its deal with Nikola, eyes a new hydrogen truck project. 2021-01-27 [cited 2023 20 August]. https://www.theverge.com/2021/1/27/22251582/gm-hydrogen-truck-navistar-oneh2-range-date

Verheul B (2019) Overview of hydrogen and fuel cell developments in China. Holland Innovation Network China

Warwick N et al (2022) Atmospheric implications of increased hydrogen use. Department for Business, Energy and Industrial Strategy

Williams D (2022) Kawasaki reveals electric and hydrogen-powered motorcycles. 2022-11-12 [cited 2023 19 August]. https://ultimatemotorcycling.com/2022/11/11/kawasaki-reveals-electric-and-hydrogen-powered-motorcycles/

Wu D (2019) Why are there so many scooters in Taiwan? 2019-08-19 [cited 2019 18 October]. https://davidwu1.medium.com/why-are-there-so-many-scooters-in-taiwan-15bbeb5c77e6

Yang C-J (2010) Launching strategy for electric vehicles: lessons from China and Taiwan. Technol Forecast Soc Change 77(5):831–834

Zemo Partnership (2023) Areas of operation—low carbon emission buses—lowCVP. [cited 2023 19 August]. https://www.zemo.org.uk/work-with-us/buses-coaches/low-emission-buses/areas-of-operation.htm

ZeroAvia (2023) ZeroAvia. [cited 2023 20 August]. https://zeroavia.com/

Chapter 4
Hydrogen Around the World

Abstract Hydrogen is emerging as an important energy option simultaneously in many regions of the world. This chapter briefly summarises diverse global perspectives and policies, particularly those in North America and Europe (the United Kingdom and the European Union). It is suggested that hydrogen for heavy duty road transport might emerge first in North America. Hydrogen for use in industrial processes (possibly linking to domestic heating) might emerge first in the UK.

In preparing this overview of hydrogen developments around the world we note and acknowledge a 2022 report on a similar theme from the International Renewable Energy Agency (IRENA 2022). The IRENA study unsurprisingly has a focus on Green Hydrogen given the organisation's subject matter expertise in that area. In our project, we give more voice to Blue Hydrogen concepts, reflecting perhaps greater enthusiasm among industrial stakeholders in Blue Hydrogen and related developments than tends to be seen coming from both academics and policy-makers. It is in the area of Blue Hydrogen that there can be expected to be the greatest benefit and insight from an industry-academia dialogue project at this time. Nevertheless, we note and concur with, the key findings from the IRENA report including:

> *Hydrogen is part of a much bigger energy transition picture, and its development and deployment strategies should not be considered in isolation.*

A second major report was issued as the 2023 U.S. National Clean Hydrogen Strategy and Roadmap (Energy Gov 2023). Given the new Inflation Reduction Act in the U.S. and its 45 V production tax credits for hydrogen, based on CO_2 footprint, Green Hydrogen is more strongly incentivized relative to Blue Hydrogen for commercial projects being proposed or undertaken. This factors into industrial considerations of the innovation and technology deployment pathway for current investments.

With such considerations in mind, that in this chapter we present observations from various parts of the world.

W. J. Nuttall et al., *Insights into the New Hydrogen Economy*,
https://doi.org/10.1007/978-3-031-71833-5_4

4.1   USA and Canada

Hydrogen is an established commodity in the USA, particularly in Texas where an established hydrogen supply infrastructure and customer base exists. As (The Hydrogen Council 2020) notes, hydrogen is seen as very important in the USA for reducing the carbon intensity of difficult-to-decarbonise sectors, particularly domestic and industrial heating, mass transit and haulage. The Fuel Cell and Hydrogen Energy Association proposes an ambitious roadmap (The Fuel Cell and Hydrogen Energy Association) where hydrogen would supply 1% of final energy demand by 2030 and 14% by 2050 using a combination of both Green and Blue Hydrogen. The key near-future market sectors are expected to be mass transit and haulage. It is worth noting that the USA has significant existing grey hydrogen production capacity (Sun, et al. 2019) and that, as such, there is significant scope for conversion to Blue Hydrogen, as demonstrated at the Port Arthur SMR plant discussed in Sect. 4.1.2. Around 1600 miles of dedicated H2 pipeline are operating in the USA with many of them located in the Gulf Coast where large hydrogen users (e.g. refineries, chemical plants) are also placed (Office of Energy Efficiency & Renewable Energy 2022). The existing and extended natural gas pipelines in the USA (c. 3 million miles) (Strategy 2020) could facilitate and expand the transportation of H_2 (i.e. blended or pure, depending on advances in technology).

4.1.1   US Federal Incentives

Arguably the origins of US support for CCUS lie with a sequestration tax credit known as 45Q (in reference to Sect. 45Q of the Internal Revenue Code). Its aim was to encourage investments in carbon capture and storage technologies in the USA and to reduce emissions mainly from natural gas, coal and other industries. The Federal level tax credit was introduced in the Energy Improvements and Extension Act of 2008 and then extended in the Bipartisan Budget Act of 2018 with final regulations set in January 2021.

45Q was initially designed exclusively for the sequestration of fossil fuel CO_2 emissions (i.e. for the geological storage of CO_2 or for its utilisation in enhanced oil recovery - EOR). In 2018, the scope was expanded, the original 75 million tonne cap was removed, and credits for direct air capture and CO_2 utilisation were allowed (IEA 2008). Geological sequestration or storage involves injecting CO_2 into underground geological formations, It can be injected and stored permanently as CO_2 or chemically transformed. A different way to sequester CO_2 is to inject it underground as part of enhanced oil recovery, also referred to as EOR.

The 45Q policy was boosted in 2018 by the Trump administration. The philosophy was that the amount of credit depended on different factors, including when the facility is placed in service and the way that CO_2 is used. The original 45Q provided annual credits (per mt of CO_2) up to $50 (for geological storage) and up to $35 (for other uses) by 2026 (Internal Revenue Service 2020). In the original legislation, after 2026 the amounts paid are to be adjusted for inflation.

According to the Global CCS Institute (Global CCS Institute 2022) the Biden administration measures to strengthen 45Q have included:

- Pushing back the acceptable latest start date of construction from 1 January 2026, to 1 January 2033.
- Lowering the capture quantity thresholds for inclusion in the policy

  - Direct Air Capture - 1000 tonnes
  - Electricity generating facility - 18,750 tonnes – and subject to further specific conditions
  - Any other industrial facility - 12,500 tonnes

- Increasing the 45Q credit value

  - Industrial and power facilities - $85/tonne (up from $50 for the original 45Q regime as applied to 2026)
  - Enhanced oil recovery - $60/tonne (up from $35 for the original 45Q regime as applied to 2026)
  - Direct Air Capture

    - CCS - $180/tonne
    - CCUS - $130/tonne

The Global CCS Institute (Global CCS Institute 2022) also observe that the Inflation Reduction Act (IRA) includes new direct pay provisions associated with 45Q. Commercial projects will receive direct payments for the first 5 years after carbon capture equipment comes into use. For non-profit organisations the equivalent support lasts longer and covers the whole 12 years of the policy.

The 45Q policy and its follow-on developments have been successful. With the levels of support outlined above, it has attracted new investments in carbon capture projects from both government and industry, especially after improvements were made to the regulation in 2018. For example, more than 40 new projects were announced between 2018 and 2021 in 12 different states, led by Texas, Illinois and Nebraska. More than 50% of the projects were concentrated on the industrial sector. There is also an important preference for dedicated geologic storage (Internal Revenue Service 2020; Congressional Research Service 2023; Clean Air Task Force 2021).

According to CCS database of the National Energy Technology Laboratory (NETL), there are around 30 active CCS projects in the USA, out of 93 active worldwide (National Energy Technology Laboratory 2023). However, and in contrast with emission intensive industries, coal and gas-fired power plants are among the least represented in the adoption of CCUS. Among the main challenges that power plants face is the low concentration of CO_2 in the emissions stream. The development of new technologies would help to increase the capture rates (currently approx. 90% for smokestack emissions).

Nowadays (2023) there are only two commercial power plants with CCUS facilities operating (Petra Nova in Texas-US, and Boundary Dam in Saskatchewan in Canada), but there are at least 20 in development worldwide (International Energy Agency 2023a). Of these, in the US alone, there are at least 10 large-scale CCUS power generation projects in advanced development, many of them to be used for EOR (Energy Ventures Analysis 2020). However, Petra Nova was temporarily closed down after drop in oil price could no longer support oil production by CO_2 EOR. Figure 4.1 shows the different CCUS projects in the US represented according to type and scale.

One can observe US government support for CCUS innovation has been both relatively consistent and relatively bipartisan, at least in comparison to other aspects of American political life. US policy for CCUS is arguably one of the major global drivers for what could become a global low-carbon hydrogen industry. A 2019 a National Petroleum Council report (National Petroleum Council 2019) provided cost curves for CO_2 capture and storage demonstrating that CO_2 capture from natural gas production (e.g. for export of LNG to Europe), methane steam reforming, ammonia synthesis, and fermentation off-gas from bioethanol production, can be fully profitable. The latter offers the potential for attractive pricing and negative emissions (Kansas Reflector 2023). The primary risk however lies in public resistance to deployment, and management of storage risks such as CO_2 pipeline leaks and induced seismicity. The latter is minimized via storage in depleted hydrocarbon reservoirs, where geology is well characterized. This relates closely to EOR, and similar, experience.

Previously, the US held about half of the global CO_2 storage projects (IEA 2008), and the new credits have further expanded US investment potential. 45Q expansion

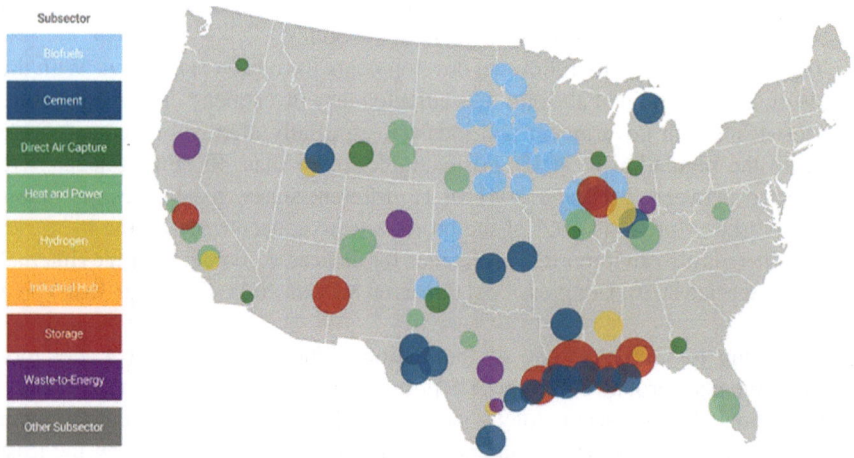

Fig. 4.1 Representation of US CCUS projects by type and scale, derived from the Clean Air Task Force's global CCUS project map, available at (Clean Air Task Force 2023) Source: Clean Air Task Force with kind permission. All rights reserved

can be thus expected to give a significant boost to US Blue Hydrogen production given its reliance on effective CCUS.

As described in Chapter 2, the Inflation Reduction Act (IRA) introduced and passed during the Biden presidency has been a game changer for hydrogen investment in the U.S, with subsidies up to \$3/kg for hydrogen production at reduced CO_2 footprint relative to current hydrogen production by steam methane reforming of natural gas. IRA includes the provision of \$8 billion of US federal funds to create a series of regional hydrogen hubs (H2Hubs) based on renewable energy sources, natural gas with CCUS and nuclear energy (Energy.Gov 2022).

Let us now examine some specific hydrogen developments from across North America.

4.1.2 Gulf Coast

4.1.2.1 Port Arthur SMR Plant

Air Products operates SMR plants at Port Arthur, Texas and these form an important part of the Texan hydrogen infrastructure. In 2013, these plants finished construction of carbon capture systems (US Department of Energy 2013). These systems are an important example of CCU, as the CO_2 captured is piped to a nearby oil field where it is utilised for enhanced oil recovery. The Port Arthur plants are only part of a significant hydrogen production infrastructure in Texas and demonstrate the practicability of CCU in the region, such developments have the potential to lead to significant emissions reduction and the expansion of Blue Hydrogen in the region.

4.1.2.2 ExxonMobil Developments

In November 2021 the Bureau of Ocean Energy Management (BOEM) announced the Gulf of Mexico Lease Sale 257 results, generating around \$191 million from bids and covering 1.7 million acres in federal waters (Bureau of Ocean Energy Management 2021). US energy giant ExxonMobil was awarded 100 shallow water blocks along the Texas coast. It is expected that the company will use these blocks for deploying CCS in Texas and to capture and store CO_2 emissions from refineries, petrochemical plants and other industries across the Houston Ship Channel (S&P Global 2021a). The expected investment for the giant hub, the world's biggest CCS project, would be around \$100 billion (with private and government funding) capturing circa 50 million tons/year of CO_2 by 2030 and twice this amount by 2040 (Bloomberg 2021).

On 22 March 2024 ExxonMobil announced a prospective deal with Japanese energy company JERA. The deal relies on US tax credit support, but is intended to supply JERA with half a million tonnes of 'Blue Ammonia' annually from 2028 (Hydrogeninsight 2024). This arrangement would represent roughly half the production from Exxon's Baytown project east of the city of Houston, Texas. This would appear to be a major step forward for the internationalization of the US Gulf Coast hydrogen economy.

4.1.3 California

While Texas and the Gulf Coast may have a claim for leading US Hydrogen thinking in terms of industrial strategy and realignment to a low carbon energy molecules future, California can surely claim to lead the US in terms of policy pressure to deploy hydrogen FCEV technologies at scale in ways to be experienced by citizens, for business and leisure. California has taken measures in terms of policy and infrastructure support to help ensure its status a leading region for US hydrogen developments. The state is home to an important Hydrogen Refuelling Network (California Energy Commission 2019). There is a network of 69 light duty refuelling stations, and 6 heavy duty stations (California Energy Commission 2023).

Despite California's contender status as the leading US state in the hydrogen energy transformation story, it is noteworthy, however, that South Korea overtook California in Hydrogen Fuel Cell electric vehicles (15,000 each) in 2022 (IEA 2023), with China placing third for a global fleet of 72,000 hydrogen FCEV.

California, however, most definitely remains a key place to watch. Both the uptake of FCEVs and construction of filling stations is set to increase significantly during the first half of the 2020s. Hydrogen in California is primarily supplied by electrolysis, consistent with Green Hydrogen. This is partially facilitated by the significant penetration of renewables in California's power grid. During 2020, California's renewable supply have seen curtailments approaching 30% of the power generated and hydrogen production is seen as a means to store energy and avoiding these curtailments in the future. Ultimately, however, hydrogen produced from curtailed renewable supply will not be sufficient to supply a significant expansion of the FCEV network. One consequence of this reality could be a shift to dedicated renewable-supplied electrolysis plants.

Perhaps the most interesting feature of the hydrogen market in California is that it already has significant policy support; there is set public funding of the expansion of the hydrogen filling enshrined in law, until a certain number of filling stations have been built. The focus of California Energy Commission (2019) is on the expansion of FCEVs into taxis and ride hailing services and the expansion of the filling network. California is an important example market where a serendipitous meeting of primary energy supply and customer has occurred, and where policy support has been an enabling factor in the expansion of the supporting infrastructure.

The deployment of hydrogen infrastructure in transport is promoted in California via different incentives and rules, such as the Low Carbon Fuel Standard (LCFS) and the Advanced Clean Trucks (ACT) regulation. The aim of the LCFS (part of the AB 32 measures to reduce GHG emissions) is to incentive the use of cleaner low carbon transportation fuels. The LCFS regulation in California was approved in 2009 and has been amended several times over the years. The LCFS standards are defined based on the "carbon intensity" (CI) of gasoline and diesel fuels and their substitutes. The standard assesses the life cycle of the GHG emissions related to the production, transport, and use of a specific fuel, including direct emissions and indirect effects. The CI scores associated with each fuel are compared to a CI benchmark for each year, and this benchmark reduces over time, in agreement with California's 2030 GHG target (20%). If the CI score is below the benchmark, it generates credits, otherwise it generates deficits, with both denominated in metric tonnes of GHG emissions. In case of deficits, credits can be used, or acquired, from another party in order to meet the annual compliance obligation (California Air Resources Board 2022). Since 2018 credits for zero emission vehicle infrastructure (i.e. capacity of hydrogen station or EV fast charging site net of actual fuel dispensed) have been included in the LCFS. The value of market for credit transactions was around \$2 billion in 2018 (California Air Resources Board 2022). On the other hand, ACT regulation encourages the adoption of zero-emission trucks by setting sales requirements with a gradual increase. Around 50% of new trucks sold in California are required to be zero-emission by 2035 (California Air Resources Board 2021). Subsidy and price support by vehicle OEM's has been required to support market growth, as hydrogen fuel costs have ranged to above \$20/kg for early stages of supply chain development (Hydrogeninsight 2022).

In Chapter 3 we discussed the somewhat difficult history of Nikola Motor as it seeks to develop hydrogen fuel cell trucks for the US market. A key part of the Nikola Motor strategy has been the development of its underpinning hydrogen supply business. Launched under the Hyla brand in January 2023, Nikola's hydrogen supply ambitions include a goal of establishing 60 large-scale hydrogen supply stations by 2026 (Fletcher 2023). Unsurprisingly the initial focus will be on California with one of the first sites being near the Port of Long Beach. Hyla intends to fuel both its own trucks and FCEVs made by other companies. Hyla reports that it will have access to up to 300 metric tonnes of fuel cell quality hydrogen per day.

4.1.4 Appalachia Blue Hydrogen Cluster

In February 2022, major energy companies including Shell, General Electric Gas power, EQT Corporation, Equinor, Mitsubishi, US Steel and Marathon Petroleum announced plans to create a Blue Hydrogen hub in Ohio, Pennsylvania and West Virginia (E&E News 2022) with carbon captured and stored underground. Blue Hydrogen will be produced using the large region's natural gas resources, exceeding Texas' natural gas production (S&P Global 2021b) (approx. 27% of USA market share in 2022 (U.S. Energy Information Administration 2023)). Many of these companies are already operating in these regions with existing natural gas facilities, hydrogen or CO_2 storage projects.

4.1.5 Alberta Carbon Trunk Line

The Alberta Carbon Trunk Line became fully operational in June 2020 and took more than a decade to be completed. It is owned and operated by Enhance Energy and Wolf Midstream and is described as the world's largest carbon capture, utilisation and storage (CCUS) system. Industrial emissions are captured and CO_2 is transported to Central Alberta and depleted into oil and gas reservoirs for use in enhanced oil recovery and permanent storage (Enhane 2023). The open access pipeline runs from the Alberta Industrial Heartland to Clive. It consists of a 240 km pipeline that can transport up to 14.6 million tonnes of CO_2 per year and required an investment of around $CDN 1.2 billion (Major Projects Alberta 2022).

4.1.6 Potential North American Leadership in Heavy Road Transport with Hydrogen

From the workshops we gained an impression that North America could be the place to lead the world in low-carbon hydrogen-fuelled trucking. The conclusion is not a consequence of narrow thinking of fuelling costs nor from a well-to-wheels analysis of GHG emissions. It comes from a wider sense of the issues facing practical road haulage in such a geographical context. Figure 4.2 shows a typical American diesel-fuelled "big rig" – a Kenworth W900. What will be the low carbon successor to such a vehicle?

Fig. 4.2 Kenworth W900 Source: PRA (Wikimedia) CC-BY-2.5

We said that our considerations would not be led by narrow economic analysis, but nevertheless let us start with an important economic concept – the total cost of ownership. ICCT has published a recent review of Total Cost of Ownership (TCO) for long haul trucks in the US. The report indicates hydrogen fuel cell and internal combustion vehicle trucks will struggle to be cost competitive with diesel. Battery electric long-haul trucks are expected to be more cost effective than diesel or hydrogen fuel cell options for more than 65% of the long-haul duty cycles. Hydrogen fuel cell truck market may be advantaged in TCO only for a small segment of the zero emission vehicle duty cycles, which brings into question whether quantities of truck production can achieve economies of scale. New battery-swapping technologies practiced now in China could further undercut the advantage of hydrogen FCEV for long distance, heavy duty trucking (International Council on Clean Transportation 2023).

Let us imagine, however, a scenario of the near future. Small high value packages (such as a new smartphone or designer sports shoes) are needed by a shopping mall outside Chicago Illinois. The items are manufactured in China and branding is very important. The items are to be sold in an upmarket location with the same retail branding as the products. It is very important to the brand that the supply chain is ultra-low carbon. The shipment will arrive at the Port of Long Beach and it needs to get to Chicago with the same supply chain efficiency that we see today. To achieve that the truck driver wants to be driving from Moab, UT to Kearney NE on the second day behind the wheel. This route will have the truck crossing the Rocky Mountains.

Points to note:

- Imagine, for comparison, a Kenworth W900 – 525 hp (386 kW) tractor for tractor-trailer haulage.
- USA drivers can drive 11 h as opposed to 9 h in Europe.
- A Big Rig on US roads has a maximum legal weight of 80,000 pounds (36.3 tonnes) of which roughly ¾ is load, i.e. roughly 25 tonnes. Just to note the maximum weight of a 40' ISO container is 20 tonnes.

A hydrogen fuelled truck can stop every 3.5 h for a 15-min refuel and driver comfort break. Hydrogen carry weight is low. Although especially for cryogenic LH2 the fuel tank itself might be quite heavy.

The key business proposition for hydrogen relates to asset utilisation (fraction of the time the vehicle is rolling) and the efficiency of the company's wage bill for drivers. We should note that asset utilisation motivated the success of hydrogen usage in large warehouses where hydrogen forklift trucks are now commonplace.

As an alternative to hydrogen, let us consider a battery electric truck, but such a truck would need to run for 11 h. Long refuel stops are not economically viable and repeated fast charging is not yet technically possible. We assume an average power rating of 200 kW for 11 h of arduous US driving.

- 200 kW for 11 h is 7.9 MJ
- The original Tesla S pack had a capacity of 100 kWh (360 kJ) – That's the original technology, we concede: it is now improved.

- For an 11-h straight drive the truck would need 22 such packs at a total weight of 11.8 tonnes (540 kg*22)
- 12 tonnes is almost half the possible cargo weight. It is simply not a practical proposition.

It is hoped that hydrogen, like diesel, might be refilled in a 15-min break at a US truck stop. To that one might say, "but in the future the electrical charging of batteries might also be quick … perhaps as quick as filling with diesel today". Well, we suggest that, if true, it would imply something else that we should consider.

US truck stops are sometimes in the middle of nowhere and they are supplying a very large amount of energy. If diesel today, or hydrogen in the future, that energy can be brought in easily by tanker.

One litre of diesel is 38 MJ. The US permits 'truck pumps' to pump 130 L per minute each, in contrast cars are filled at no more than 50 L per minute. For trucks it's 2.16 L per second. Roughly 80 MW. Imagine a US 18-bay truck stop running at heavy utilisation. We can admit that might be a very big truck stop, but it helps to make our point. Such a facility might have 12 pumps dispensing at any one time. Such a truck stop would be dispensing nearly 1GW at that busy moment (984 MW). With thanks to Vaclav Smil for his book Power Density for inspiration on this point.

Now imagine if we wanted to meet that nearly 1GW of peak refuel demand with electricity, a spur from the grid might well be impractical … perhaps a nuclear plant at the truck stop? At this point we suggest that bringing in hydrogen, or some other low carbon synthetic fuel by road, starts to look positively attractive.

Lastly, it must always be remembered that hydrogen vehicles are electric vehicles. Therefore, most innovation in BEV is innovation for FCEV. Once the initial BEV products have plateaued in the market, it should then be expected that Hydrogen vehicles should get a lot cheaper quite quickly, as technology is pushed further in a search for expanded market share. We note with interest that Nikola Motors first entered the market with a BEV offering before shifting focus to FCEV technology.

We note that in Germany in September 2023 it was announced that a Daimler hydrogen fuel cell truck had demonstrated its ability to travel more than 1000 km on a single fuelling of hydrogen (Daimler Truck 2023). We posit that such demonstrations relate directly to the beneficial role that hydrogen can play in heavy road transport. We suggest that such capabilities will be especially important in North America, and perhaps other relatively low-density developed economies, such as Australia.

In this section we have referred to several companies developing hydrogen trucks. In fact, the number of such companies worldwide is large. Another example would be VDL Group of the Netherlands. It has recently entered into a partnership arrangement with Toyota (Manthey 2023). We list these developments in our section devoted the USA given our sense that North America could be a particularly important early market for such innovations.

4.2 UK

The UK has long had a particular interest in hydrogen production and use. With its extensive natural gas grid, hydrogen presents a significant opportunity for decarbonisation of home heating; a difficult end-user service to decarbonise. Prime Minister Boris Johnson's UK infrastructure speech of June 2020 mentioned hydrogen production specifically as a measure which would enable reductions in CO_2 emissions:

We lead the world in quantum computing, in life sciences, in genomics, in artificial intelligence, space satellites, net zero planes and the long-term solutions to global warming: wind, solar, hydrogen technology, carbon capture and storage, nuclear, and as part of our mission to reach net zero CO_2 emissions by 2050 we should set ourselves the goal, now, of producing the world's first zero emission long-haul passenger plane - 'Jet Zero' - let's do it! – Boris Johnson (YouTube 2020)

As noted earlier, in December 2020, the UK government announced its Ten Point Plan (HM Government 2020) and the aim to develop (along with industry partners) 5GW of low carbon hydrogen production capacity and to capture 10Mt of CO_2 per year by 2030. Among the specific schemes to support this are the creation of £240 million Net Nero Hydrogen Fund and the allocation of £1 billion to facilitate the deployment of CCUS in four industrial clusters or "Super Places" (2 sites by mid 2020s and the 4 sites by 2030). The H2 carbon strategy was realised in October 2021 and set the key steps and foundations to achieve the UK hydrogen economy with strong support from the industrial sector. The carbon strategy linked to the hydrogen strategy published a few weeks earlier. In August 2021 the UK Government had published its hydrogen strategy (Department for Business, Energy and Industrial Strategy 2021). It laid out nine strategic goals for 2030 including:

- **Progress towards 2030 ambition**: 5GW of low carbon hydrogen production capacity with potential for rapid expansion post-2030; hope to see 1GW production capacity by 2025.
- **Decarbonisation of existing UK hydrogen supply**: Existing hydrogen supply decarbonised through CCUS and/or supplemented by electrolytic hydrogen injection.
- **Lower cost of hydrogen production**: A decrease in the cost of low carbon hydrogen production driven by learning from early projects, more mature markets, and technology innovation.
- **Preparation for ramp up beyond 2030** – on a pathway to net zero: Requisite hydrogen infrastructure and technologies are in place with potential for expansion. Well established regulatory and market framework in place.

The report observed a wide range of activity underway, or planned and announced, for clean hydrogen production as shown in Fig. 4.3 (taken from Department for Business, Energy and Industrial Strategy (2021)).

Subsequent events, including the upward trend of electricity prices and Russian President Vladimir Putin's invasion of Ukraine, led then Prime Minister, Boris

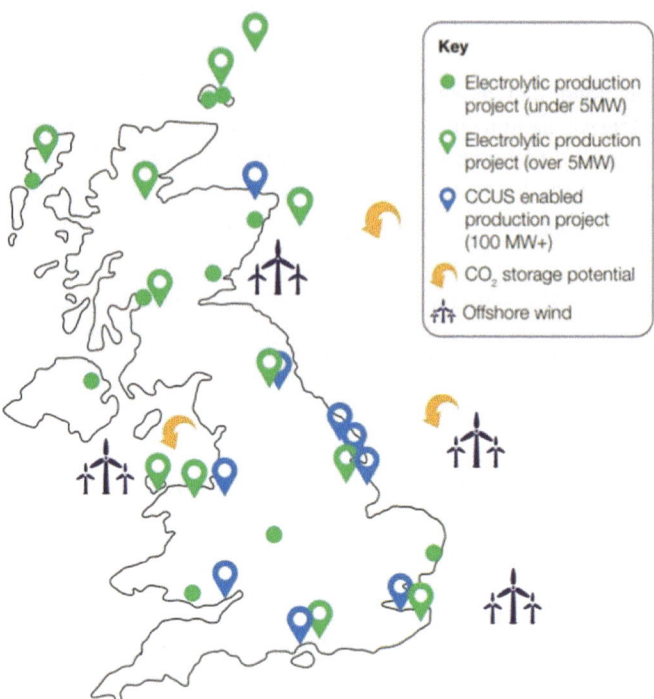

Fig. 4.3 UK Clean Hydrogen Capabilities and Disclosed Plans August 2021. Source: BEIS Hydrogen Strategy 2021 (UK government). Note: graphic excludes undeclared private sector plans

Johnson, to announce in April 2022 an Energy Security Strategy to accelerate net zero. The strategy proposes to increase the UK's ambition for low carbon hydrogen production capacity by 2030, reaching 10 GW. There is a clear indication of the role that electrolysis will play on this, suggesting that at least 50% of the target capacity be met via electrolysis technology, including green and pink hydrogen (Department for Business, Energy and Industrial Strategy 2022).

In recent years, several organisations have presented infrastructure projects that will make the first steps towards the significant development of hydrogen infrastructure in the UK. These comprise ambitious regional projects and go further in suggesting means by which they might be expanded across the entire country.

At our third event we heard challenge to the notion that UK energy decarbonization is inevitably a story of complete electrification of the economy. Arguably hydrogen could replace Natural gas in the existing plastic low pressure distribution systems. Such local distribution capabilities could be inter-connected with a radically re-structured (or new) national/international hydrogen transmission system at higher pressure. Such a scenario would involve a conversion of the UK's domestic gas supply. Marc Cochrane reminded those present that "changing gas quality has been done before. 13 million homes and 40 million appliances were converted (from 'town gas' to natural gas) in the period 1967–77".

At the first event held in 2021 Ian Hibbitt had also considered gas distribution conversion as an alternative to electrification. He suggested that producing hydrogen and electricity is not a problem in the UK, but the distribution (movement of energy in whatever form to the end user) potentially is. BOC estimated UK movement of energy in gas pipes, electricity pylons and petroleum pipes to be annual 736TWh, 275TWh and 634TWh respectively. We note that the electricity network is the smallest contributor in that list. The carrying capacity of gas pipeline infrastructures is an important consideration that must not be neglected by policy-makers when considering options for a rapid and very large-scale shift to low carbon energy distribution.

Dan Sadler, suggested in our first event that the most important part of decarbonisation is to get infrastructure that can support both Blue and Green Hydrogen projects in place. He indicated that Equinor believes that Blue Hydrogen will initially dominate the transition, and once the infrastructure is in place, the Green Hydrogen will increasingly become more and more competitive and potentially start to displace blue, over the decades to come.

Writing in 2023, we can say that the UK's east coast hydrogen cluster is taking shape around a project known as Zero Carbon Humber, entailing 35 million tonnes per annum of CO_2 emissions per industrial cluster, requiring parallel runs of pipelines for H_2 and CO_2, with use of bioenergy and negative emission to achieve net zero. The UK Climate Change Committee (CCC) says that CCS is a necessity at a minimum level of 10 MT per annum in 2026 based on at least two clusters. One, or more, of these clusters should be at a scale sufficient for making low carbon hydrogen. In October 2024 approval was given by the UK government for a major £21.7 billion funding package over 25 years to support UK CCUS including the HyNet project described in the next section.

Our meeting held in March 2021 heard that the Humber represented a most attractive location for such an industrial cluster.

It was further observed that a new autothermal natural gas reformer on the Humber rated at 600 MW and producing Blue Hydrogen should be able to operate at less than 15 g/kWh CO_2 emission; with a 95% CO_2 capture rate operating at 80% efficiency. Industrial clusters are well placed to scale supply and to match to demand. A cluster, however, needs to include hydrogen suppliers and also end users. Most importantly there needs to be an appropriate infrastructure to link these various elements.

At our 2022 London meeting we heard that there are commercially ready designs for hydrogen-fuelled 'combi' boilers from Worcester-Bosch and Baxi Heating. At the time of writing the government still hopes to be able to develop trial projects for hydrogen domestic heating despite significant public concern and the cancellation of some plans for local hydrogen distribution and use.

Looking to the bigger picture, some specific UK hydrogen initiatives are discussed below.

4.2.1 HyNet North West (HyNet 2018)

The HyNet North West project is a proposed to provide hydrogen infrastructure around the Liverpool, Manchester, and Cheshire regions. Near-pure hydrogen would be delivered to local industry and would be blended into the public natural gas grid at concentrations of up to 20%. The project uses only proven technology, and its proponents argue that it is the fastest and lowest cost option (per unit CO_2) reduced when compared to competing options for home heating. This project would use ATR as the means of hydrogen production. CO_2 from the ATR plant, as well as some already captured by existing industry in the region would be transported via pipeline and stored in a disused gas field. The project CAPEX costs are expected to be in the region of £920 m.

The project deliberately relies on existing technology and infrastructure wherever possible. Whilst the project team notes that there are no anticipated technical show-stoppers for the alternative proposition of 100% hydrogen supply, it does concede that some end-users will find it difficult, or impossible, to convert fully to hydrogen. As a result, some industry users, particularly gas turbine operators will find it easier to cope with a hydrogen-natural gas blend. It therefore argues that a hydrogen blend in the public gas grid is a prudent and pragmatic decision on the observation that industry, policymakers and regulators will not buy into a full conversion to hydrogen without a prior successful demonstration.

While blend-based approaches are relatively easily implemented and represent an important positive development in the journey to Net Zero, at some point in the medium-term future a further technological shift (such as to pure hydrogen distribution or to electrification) would be required. Hence the issue soon becomes whether current developments in the direct of hydrogen blends represent a risk of technological lock-in that could make deeper decarbonisation more difficult than would otherwise have been the case. Examples of alternative strategies include, for example, a single-step shift from natural gas to 100% hydrogen distribution without delay – or a bold shift to total electrification. Both these other options are more radical and hence more prone to risk than the more incrementalist ideas of hydrogen-natural gas blending. One can argue that neither of these alternative and more ambitious strategies is currently credible. With that observation in mind, one can ask oneself whether pragmatic progress is better than bold failure.

4.2.2 H21 Leeds City Gate (H21 2019)

The H21 Leeds City Gate project has pioneered the ambitious idea of converting the entirety of the public natural gas grid in the cities of Leeds and Hull to pure hydrogen. The project asserts that the region is capable of being converted incrementally over successive summers when gas use will be at its lowest. In this case, SMRs would be used to produce hydrogen. CO_2 from the SMR plants should be captured and stored beneath the North Sea. This project would require the full conversion of all end use combustion points and the careful checking of the distribution system. It also envisages a role for significant hydrogen storage. Salt caverns within the region are proposed for both short-term and inter-seasonal storage. Whilst this project relies on existing technology, it concedes that the domestic appliance market requires development and that some form of financial stimulus will be required to facilitate this. The project CAPEX costs are expected to be in the region of £2bn. As with HYNET NW the original project timeline appears to be slipping in 2023.

4.2.3 H2H Saltend (Equinor 2020)

Hydrogen to Humber (H2H) Saltend project has been led by Equinor and is part of the Zero Carbon Humber Hub partnership. The H2H Saltend project, placed within the East Coast Cluster, has passed the eligibility criteria for Phase-2 cluster sequencing process. H2H Saltend aims to develop a 600 MW low carbon H2 production plant with carbon capture able to reduce the industrial site emissions by around 1 million tonnes per year (equivalent to 30% reduction in the site's total carbon emissions) (equinor 2020).

The Saltend chemicals park is the UK's largest industrial emitter. It does, however, have proximity to both the port infrastructure at Hull, and the Endurance aquifer, an extensive potential CO_2 store in the North Sea. The initial stages of the project would see the installation of ATR hydrogen production plants. The already integrated nature of the infrastructure at Saltend means that industrial customers would be able to benefit from the hydrogen produced by the ATR plants and be able to utilise the associated CCS facilities. The project proposal references the H21 Leeds City Gate project as potentially synergistic and states that the H2H project is specifically designed to be expandable. It even proposes that hydrogen could be exported, and that CO_2 storage infrastructure could generate revenue by importing CO_2 by ship.

H2H Saltend benefits from the participation of the Norwegian energy company Equinor which is highly experienced in matters relating to CCS. Equinor is currently working with Shell and Total on the Northern Lights project. The first carbon storage project on the Norwegian Continental Shelf (Equinor 2023).

4.2.4 Midlands Engine Partnership

In December 2021 the English regional partnership known as the Midlands Engine Partnership (MEP) published its hydrogen technologies strategy (Midlands Engine and Green Growth 2021). The MEP Chairman, Sir John Peace, observed *"The Midlands is the UK's manufacturing heartland and one of the most important national locations for hydrogen innovation and application. We have a unique blend of diverse manufacturing ingenuity and capability, world-leading commercial, indus-trial and academic innovation, and ideal geographic location"* (Midlands Engine and Green Growth 2021). We understand that by February 2022 the MEP was observing that its hydrogen technologies strategy had the potential to generate, or safeguard, 167,000 jobs, to contribute £10 billion (GVA) to the economy, yield a 17 million Tonne reduction in CO_2 emissions (29%). The key to such success would be the need to secure necessary investment (Kendall 2023).

4.3   Continental Europe

Europe also has significant interest in hydrogen production. The region benefits from established international regulatory structures and a significant natural gas distribu-tion network that could facilitate the distribution of hydrogen. It also has significant import/export capacity and there is a great deal of interest in hydrogen production from key ports. The earliest end-users for hydrogen are expected to be part of trans-portation, particularly mass transit and haulage (International Energy Agency 2019), as well as the displacement of natural gas in the existing grid and for supply to industry (Fuel Cells and Hydrogen 2 Joint Undertaking 2019).

This section details some of the proposed projects and national strategies that could contribute to the expansion of hydrogen in Europe.

4.3.1   Port of Rotterdam

The port of Rotterdam is an important energy import/export hub (Port of Rotterdam 2020). The port authority notes that the displacement of the energy throughput of Rotterdam with Green Hydrogen would require significant expansion renewable generation and electrolysis domestically, as well as significant import of Green Hydrogen from regions of the world where renewable generation is significantly cheaper. The report therefore also notes that Blue Hydrogen could also play a signif-icant role and that consistent with such thinking existing Grey Hydrogen plants. will need retrofitting with CCS for conversion to Blue Hydrogen production. Overall, the report expects that hydrogen will reach cost competitiveness with fossil fuels in the early 2030s and proposes an ambitious expansion of the port's infrastructure to cater for the increased hydrogen throughput this would entail.

4.3.2   NortH2

The Netherlands' NortH2 project is a recently proposed collaboration between Shell, domestic natural gas supplier Gasunie, and Groningen Seaports to develop a Green

Hydrogen infrastructure and FCEV fuelling infrastructure (H2 View 2020). The project would see the construction of significant electrolysis capacity supplied by a dedicated wind farm for the generation of Green Hydrogen (greentechmedia 2020). The resulting hydrogen would be supplied to filling stations via Gasunie's natural gas infrastructure.

4.3.3 Hyport

HYPORT is a Green Hydrogen project slated for the Belgian port of Oostende (DEME Group 2020). This project will see the installation of electrolysis capacity in proximity to the port with the intention of using peak renewable wind energy to produce hydrogen at times when supply exceeds demand. The product will be supplied to end users across electricity, transport, and industrial sectors. The timescale for this project is ambitious with power generation capacity running on Green Hydrogen produced by a pilot electrolysis plant installed by 2022. The final hydrogen production plant is expected to be completed shortly thereafter in 2025.

4.3.4 Hydrogen in Germany

Germany published a national hydrogen strategy in 2020 (The Federal Government 2020). In the near term, Germany expects hydrogen to be important for decarbonisation of the industrial chemicals and steelmaking sectors, as well as seeing some uptake in transportation. In the longer term, it is expected to play a more important role in transportation, particularly haulage, industrial heavy machinery, and mass transit where it is viewed as being complementary with battery technologies. The strategy asserts the importance of hydrogen and related synthetic fuels in decarbonisation of both the heating network, which is primarily supplied by natural gas, and aviation and maritime transportation.

Blue Hydrogen is conspicuous in its absence from the German strategy document. The focus for domestic production is on the expansion of renewable electricity generators and electrolysis for Green Hydrogen production. The report does, however, concede that this will only be sufficient for a small portion of the expected hydrogen demand in Germany and that there will be significant reliance on imported hydrogen and its related products. It would seem likely, then, that Blue Hydrogen will play an important part in Germany's future hydrogen economy even if it is not part of domestic production.

Germany has also published a detailed analysis of the infrastructures required for widespread conversion of transportation to either BEVs or FCEVs (Robiniusa 2018). This detailed and comprehensive comparison of infrastructure requirements argues that hydrogen and electrification should both pay a role in decarbonisation of the transportation sector, indeed arguing that the synergies between hydrogen for transportation and for the wider industry offer unique opportunities for multi-sector decarbonisation.

4.3.5 Wider EU Considerations

Austria, Belgium, France, Italy and Norway have also made inroads towards hydrogen production for heating, transportation and shipping (International Energy Agency 2019). Spain and Portugal have active programs linking to the use of offshore wind (Partidário et al. 2020).

Again, across the EU policy is focussed on the potential for Green Hydrogen linking to ambitions for renewables-based electrification. Looking beyond such long-held positions, a key issue across the European Union in recent years has been policy progress around a wider "taxonomy" through which the EU seeks to define sustainable activities (European Commission 2023a; Doyle 2023). Two technologies proved particularly controversial during the debate around the proposed policy. They were CCUS and nuclear power. Both were eventually admitted into the taxonomy albeit with caveats and restrictions. In October 2023 the Oxford Institute for Energy studies issued a report with the title: *Carbon Capture Usage and Storage the new driver of the EU Decarbonization Plan?* (Borchardt 2023). The report draws parallels between that 2023 initiative and the EU's earlier Hydrogen Strategy published in 2020 (European Commission 2023b).

The OIES notes: "*Similar to the development of a hydrogen backbone, a predictable and transparent regulatory framework for the future CO_2 transport infrastructure should be developed ensuring third-party access and including a CO_2 infrastructure network operator, network planning and regional cooperation*" (Borchardt 2023).

The 2020 EU Hydrogen Strategy was accompanied by measures for a new directive on common rules for the internal markets in renewable and natural gases and in hydrogen. Interestingly that directive admitted a role for both Blue and Green hydrogen in the EU energy system of the future (EUR-Lex 2023).

4.4 Asia Pacific

4.4.1 Japan

Japan has a significant interest in hydrogen which will play a crucial role in the decarbonisation of its economy especially its industry (Japan is one of the most industrialised countries worldwide) as well as other sectors including transport, heating and power. Japan's Basic Hydrogen Strategy (Ministerial Council on Renewable Energy Hydrogen and Related Issues 2017) issued in 2017, outlines Japan's diverse approach to developing a hydrogen economy. Investment has been committed to developing hydrogen for power generation, mobility, and industrial heating requirements. Supporting the broader strategy is the Strategic Roadmap for Hydrogen and Fuel Cells, updated in March 2019, which encourages the development of infrastructure for hydrogen-based transportation (Hydrogen and Fuel Cell Strategy Council 2019). The Roadmap sets new targets on the specifications of technologies including cost breakdowns and the measures required to meet them. Japan is a leader in full cell technology and aims to be one of the largest global exporters of such technology. The development of the hydrogen supply chain by 2030, which would contribute to increasing the scale of production and a lowering of costs, is also part of the agenda. This would require enhancing and building government-level relationships with countries with important renewable sources. For instance, in April 2022 Japan and Germany agreed to launch a governmental consultation to cooperate on hydrogen as an alternative to gas and coal (Hydrogen Central 2022a).

It would seem that with the decline of nuclear power in Japan following the Fukushima-Daiichi accident that the Japanese government and country's academia considers hydrogen one of the most important routes to national decarbonisation.

4.4.2 China

China has been investing heavily in the hydrogen market for years, with hydrogen buses and more recent commitments to expand the development and adoption of FCEVs. Whilst the Chinese government's approach to investment and regulation has seen some criticism (Verheul 2019) in the past, 2019 saw it renew its commitment to the rollout of hydrogen transportation with a $17bn investment in the industry (News 2019). Hydrogen is expected to play an important role in China's decarbonisation economy by 2060, with the value of hydrogen industry estimated at $154 billion in the late 2020s and up to 12 times that by 2050. Among the key industrial giants with Green Hydrogen projects are Ningxia Baofeng Energy Group and China Baowu Steel Group (Murtaugh 2023).

Recent developments driven by the Chinese government are observed in some regions such as Inner Mongolia, with several policies issued in July 2021 that encourage the development of a hydrogen energy industry, including Green Hydrogen produced from wind and solar power linking to energy storage (Hydrogen Energy Industry Promotion Association 2023). For instance, the Inner Mongolia Green Hydrogen project to be implemented in the cities of Ordos and Baotou and to be operational in middle 2023, foresees to install wind (1870 MW) and solar (370 MW) power capacity to produce 66,900 tons of renewable hydrogen per year. 20% of the electrical capacity is expected to be exported to the grid, while the rest will produce the Green Hydrogen. At least 450 MW of electrolysers would be required for the project, equivalent to more than twice the global market in 2020 (Petrova 2023).

China is currently third in deployment of Hydrogen FCEV, behind South Korea and California. However, China deploys 95% of the world FCEV trucks and 85% of FCEV buses (Hydrogeninsight 2023). Battery swapping for Electric Vehicle trucks has become competitive with rapid refuelling for hydrogen fuel cell vehicles.

4.4.3 Australia

Australia also has an interest in hydrogen and published a national roadmap in 2018 (Bruce et al. 2018). In the near term, this is primarily directed towards the decarbonisation of the transport sector, and as a means to supplying power to remote areas. Longer term, it considers broader decarbonisation of the domestic and industrial heating sectors.

Clean hydrogen is one of the priority low emissions technologies included in the Australian federal Technology Investment Roadmap launched in 2020. The goal is to produce clean hydrogen under \$2/kg (Australian Government 2023a). The development of international partnerships of low emission technologies has been acknowledged, with five new international partnerships announced in 2021. Approximately A \$566 million has been committed by the Government to these partnerships (Australian Government 2023b). Australia is preparing to be a global hydrogen exporter. There are additional schemes in place that are supporting the scale up of the hydrogen industry, among them are the A \$464 million Clean Hydrogen Industrial Hubs program (to develop up to 7 hubs in regional Australia) and funds provided by the Australian Renewable Energy National Agency (ARENA). Around A \$146 millions have been committed to hydrogen projects since 2015.

In Australia the COAG Energy Council has proposed a National Hydrogen Strategy within which the case for hydrogen is made in very clear terms, see Fig. 4.4 (taken from COAG Energy Council (2019)).

Australia is a region to watch in the emerging global hydrogen economy. Australia is a major energy exporter, with key exports including coal, oil, liquefied natural gas, and uranium. For example, Australia is the world's largest LNG exporter, overtaking Qatar in 2020, and it exported more coal (in energy terms) than any other country in 2020. It was second only to Indonesia in terms of coal exports by mass (US Energy Information Administration 2022). Australia has only a small (circa 1% of its energy industry) renewable energy industry but there are bold ambitions for growth. There are plans to greatly expand hydrogen from renewable energy, but Australia has also pioneered the production of hydrogen from Brown Coal (lignite) at its Hydrogen Energy Supply Chain pilot project based in Victoria. In addition, Australia has a lignite to syn-gas facility: the Leigh Creek Energy Demonstration Project, operating since 2019 in the Telford Coal Basin in South Australia.

In September 2024, an update to the National Hydrogen Strategy was released. One of the key objectives of this strategy is to make Australia's hydrogen industry globally cost-competitive. To support hydrogen production, two major incentive schemes have been introduced: the Hydrogen Production Tax (which offers a \$2 per

 **1 kg** of hydrogen is enough to travel up to **100 km** in a **Hyundai Nexo**

 Travelling in a **Hyundai Santa Fe** uses **7.5 L** of diesel or **9.3 L** of petrol

  Driving a **Hyundai Nexo** compared to a diesel **Hyundai Santa Fe** avoids **0.2 kg CO₂-e / km** driven or **20 kg CO₂-e per kilogram** of hydrogen used

 **1 kg** of hydrogen in a fuel cell could power a **1,400 watt** electric split-cycle air conditioner for **14.5 hours**

Replacing Australian grid electricity with electricity from **hydrogen** avoids **0.75 kg CO₂-e / kWh**, or **15 kg CO₂-e per kilogram** of hydrogen used

 **1 tonne** of **hydrogen** is equivalent to around **3.4 times** the average annual consumption of an Australian house with **gas heating**

 Replacing **natural gas** with **hydrogen** avoids **0.052 tonnes CO₂-e / GJ** of **natural gas** or **6.2 tonnes CO₂-e per tonne** of **hydrogen**

Fig. 4.4   Potential attributes of hydrogen in an Australian context Source (COAG Energy Council 2019). Source: Commonwealth of Australia Australia's National Hydrogen strategy

kilogram tax credit for eligible renewable hydrogen producers), and the Hydrogen Headstart program (designed to bridge the financial gap between the production cost and sale price of renewable hydrogen).

4.5   India

In January 2022, the Indian government approved the National Green Hydrogen Mission, which aims to make India a hub for the production and export of green hydrogen (National Portal of India 2023). India is a complex and somewhat specific space for hydrogen opportunity. Particularly relevant are the the very strong role played by coal in the Indian economy (Madhavi and Nuttall 2019) and the issues of air quality facing major Indian cities. For an interesting overview of dynamic issues in Indian air quality see reference: (IQAir 2023). One of us (WJN) intends to write with Professor Bernaurdshaw Neppolian of SRM-IST Chennai on the issues facing hydrogen developments in India. At the time of writing that separate output is currently in-press (Nuttall et al. 2025).

4.6 Gulf Cooperation Council (GCC) Countries

Hydrogen is playing a central role in the energy transition strategies in GCC states, with a set of national hydrogen strategies and projects launched recently. According to the Middle East Institute, the GCC finds hydrogen attractive not just for its green credentials but for its possibility to mirror the GCC global export patterns due to its industrial similarities with hydrocarbons (material, social and economic) (Chibani 2023). The region has a strong potential to lead in hydrogen production, with a focus on Blue and Green Hydrogen. The GCC has plenty of both hydrocarbons and renewable energy sources (The World Bank 2023).

The Saudi Arabia 2030 vision, released in 2016, aims to implement strategic reforms across all the sectors that will help to reduce their oil dependency (Kingdom of Saudi Arabia 2023). The national hydrogen industrial strategy, under development, can help to meet this mandate. Saudi Arabia wants to become a top hydrogen supplier worldwide with production targets of 2.9 Mt/yr by 2030 and 4 Mt/yr by 2035. Currently the focus is getting market share on Blue Hydrogen specifically in the form of ammonia (Hydrogen Central 2022b). NEOM, a futuristic regional development, is a key project of the Saudi Vision 2030. NEOM Green Hydrogen Co. (NGHC), a joint venture between ACWA Power, Air Products and NEOM, is building the world largest utility scale Green Hydrogen production plant with a total investment of US$8.4 billion. The hydrogen plant located in Oxagon (NEOM industrial city) requires 4 GW of wind and solar energy to produce up to 600 ton of hydrogen daily (in the form of ammonia) by the end of 2026 (Neom 2023). An exclusive 30-year offtake agreement with Air Products of the total production has been secured. More recently, ARAMCO and NEOM have signed a joint agreement to install an e-fuel demonstration plant that will produce 33 barrels per day of synthetic gasoline from both, carbon capture carbon and renewable-based hydrogen (Aramco 2023).

Oman Vision 2040 was approved in December 2020. It comprises a set of national sector strategies and calls for the diversification of energy resources with a target of a non-oil share of GDP of more than 90% by 2040 (Sultinate of Oman 2023). In August 2021, the national alliance for Green Hydrogen (Hy-Fly) was created, with the aim of place Oman on the global map of major Green Hydrogen production. The Oman Green Hydrogen Strategy, launched in October 2022, has set specific targets for boosting green hydrogen, aiming to produce more than 1 Mt/yr by 2030, 3.75 Mt/yr by 2040 and up to 8.5 Mt/yr by 2050 (International Energy Agency 2023b). Oman could become the sixth largest Green Hydrogen exporters by 2030 (International Energy Agency 2023b). A subsidiary of Energy Development, Hydrogen Oman (Hydrom), has recently signed three agreements for the development of the first green hydrogen projects, worthing over US$20 billion. The total annual production is expected to be 0.5 Mt/yr with 12 GW of renewable energy capacity (Forbes Middle East 2023).

The updated United Arab Emirates (UAE) energy strategy, aims to accelerate the energy transition by incrementing the share of clean energy (including nuclear power) in the generation mix and achieving net zero in the energy sector by 2050. By 2030 it is expected to triple the capacity of renewable energy and to achieve an emission factor of 0.27 $kg/CO_2/kWh$ (Gulf News Report 2023). The National Hydrogen Strategy approved in July 2023, aims to enhance the UAE position as a leading global producer of low-carbon hydrogen by 2031. Among its targets is the reduction of emissions in hard-to-abate sectors by 25% in 2031 and 100% in 2050. Hydrogen production is planned to reach 1.4 Mt/yr by 2031 and 15 Mt/yr by 2050. The strategy also requires the establishment of a hydrogen research and development centre and two "hydrogen oases" (fuelling stations) by 2030 (United Arab Emirates 2023). The Green Hydrogen project implemented by Dubai Electricity and Water Authority (DEWA), in collaboration with Expo 2020 Dubai and Siemens Energy, produces hydrogen using solar power. The project is testing future uses of hydrogen, together with energy generation and transportation (Dubai Electricity & Water Authority (DEWA) 2023). DEWA is also collaborating with Emirates National Oil Company (ENOC) to test the use of hydrogen in mobility, making use of the existing green hydrogen plant (ENOC 2023).

4.7 Africa

The Africa Green Hydrogen Alliance (AGHA) was announced by Kenya, South Africa, Namibia, Egypt, Morocco, and Mauritania in 2023 (Deloitte 2023). It is observed that abundant renewable solar power and large land masses with low population can provide Green Hydrogen at up to four-fold lower costs than production in Europe. The production possibilities exceed anticipated local needs and a pipeline connection to Europe is contemplated.

South Africa is a particular focus for hydrogen innovation. A key part of the context for African innovation in clean energy is the possibility of supporting funds from the developed economies of the global north. Such ideas have featured strongly in the Kyoto process. For example COP-26 conference held in Glasgow, Scotland in the autumn of 2021, resolved to provide South Africa with US $8.5bn to help it phase out coal (Argus 2021). The associated spending is to be decided by a group called the 'joint task force'. Arguably the need should be to phase down global CO_2 emissions, not necessarily coal mining, but it appears that this is a measure against coal mining irrespective of the potential for low-emission innovation, in the area of synthetic fuels with CCUS, for example.

Egypt has a strong basis for developing a Green Hydrogen industry. The country has abundant solar and wind energy, including options for pipeline transport to Europe (State Information Service 2023; The Oxford Institute for Energy Studies 2023). The green hydrogen production roadmap, released in June 2023, identifies the challenges that difficult production for local players and potential pathways to accelerate green fuel generation. Egypt has already signed framework agreements with international companies to build green hydrogen and ammonia plants in the Suez Canal Economic Zone (SCEZ), worth around US\$ 83 billion. Nine green hydrogen and ammonia plants will be built, with the aim to produce once fully operational 2.7 Mt/yr of green hydrogen and 7.6 Mt/yr of ammonia (Enterprise Climate 2023). Egypt is also evaluating a potential collaboration with Switzerland's Smartenergy to build a US\$ 1billion green hydrogen plant (Kotb 2023).

References

Aramco (2023) Aramco and ENOWA to develop first-of-its-kind e-fuel demonstration plant [cited 2023 7 November]. https://www.aramco.com/en/news-media/news/2023/aramco-and-enowa-to-develop-first-of-its-kind-efuel-demonstration-plant

Argus (2021) South Africa gets \$8.5bn to phase out coal: update [cited 2023 7 November]

Australian Government (2023a) Low emissions technology statement 2022. Department of Climate Change, Energy, the Environment and Water: Australia

Australian Government (2023b) State of hydrogen 2021. Department of Climate Change, Energy, the Environment and Water: Australia

Bloomberg (2021) Exxon floats \$100 billion federal-backed carbon capture hub [cited 2022 20 June]. https://www.bloomberg.com/news/articles/2021-04-20/exxon-floats-100-billion-govern ment-backed-carbon-capture-hub#xj4y7vzkg

Borchardt K-D (2023) Carbon capture usage and storage the new driver of the EU decarbonization plan? The Oxford Institute for Energy Studies

Bruce S et al (2018) National hydrogen roadmap. CSIRO, Australia

Bureau of Ocean Energy Management (2021) Gulf of Mexico Lease sale results announced [cited 2022 20 June]. https://www.boem.gov/newsroom/press-releases/gulf-mexico-lease-sale-results-announced

California Air Resources Board (2021) Advanced clean trucks fact sheet [cited 2022 20 June]. https://ww2.arb.ca.gov/resources/fact-sheets/advanced-clean-trucks-fact-sheet

California Air Resources Board (2022) Low carbon fuel standard [cited 2022 20 June]. https://ww2.arb.ca.gov/our-work/programs/low-carbon-fuel-standard/about

California Energy Commission (2019) Joint agency staff report on assembly bill 8: 2019 annual assessment of time and cost needed to attain 100 hydrogen refueling stations in California

California Energy Commission (2023) Hydrogen refueling stations in California. 2023 current-date [cited 2023 13 September]. https://www.energy.ca.gov/data-reports/energy-almanac/zero-emission-vehicle-and-infrastructure-statistics/hydrogen-refueling

Chibani A (2023) Hydrogen as a fuel of the future: moving the GCC's climate mitigation agenda forward [cited 2023 7 November]. https://mei.edu/publications/hydrogen-fuel-future-moving-gccs-climate-mitigation-agenda-forward

Clean Air Task Force (2021) Carbon capture in the U.S. is growing like never before, but further policy support is crucial. 2021–07–29. https://www.catf.us/2021/07/us-carbon-capture-growth/

Clean Air Task Force (2023) U.S. carbon capture project map [cited 2023 17 November]. https://www.catf.us/ccsmapus/

COAG Energy Council (2019) Australia's national hydrogen strategy. COAG Energy Council

Congressional Research Service (2023) The Section 45Q tax credit for carbon sequestration

Daimler Truck (2023) Daimler Truck #HydrogenRecordRun: Mercedes-Benz GenH2 Truck cracks 1,000 kilometer mark with one fill of liquid hydrogen [cited 2023 7 November]

Deloitte (2023) Green hydrogen: Energizing the path to net zero [cited 2023 13 September]. https://www.deloitte.com/global/en/issues/climate/green-hydrogen.html

DEME Group (2020) HYPORT: green hydrogen plant in Ostende. https://deme-group.com/news/hyportr-green-hydrogen-plant-ostend

Department for Business, Energy and Industrial Strategy (2021) UK hydrogen strategy. HM Government

Department for Business, Energy and Industrial Strategy (2022) British energy security strategy [cited 2022 7 April]. https://www.gov.uk/government/publications/british-energy-security-strategy/british-energy-security-strategy?module=inline&pgtype=article

Doyle DH (2023) A short guide to the EU's taxonomy regulation [cited 2023 7 November]. https://www.spglobal.com/esg/insights/a-short-guide-to-the-eu-s-taxonomy-regulation

Dubai Electricity & Water Authority (DEWA) (2023) Green hydrogen is one of DEWA's solutions to diversify energy sources and provide 100% of total power capacity from clean energy sources by 2050 [cited 2023 7 November]. https://www.dewa.gov.ae/en/about-us/media-publications/latest-news/2022/03/green-hydrogen

E&E News (2022) Companies plan to build CCS, hydrogen hub in Appalachia. E&E News

Energy Ventures Analysis (2020) Understanding 45Q: the carbon capture tax credit [cited 2023 19 August]. https://www.evainc.com/energy-blog/45q-the-carbon-capture-tax-credit/

Energy.Gov (2022) DOE launches Bipartisan infrastructure law's $8 billion program for clean hydrogen hubs across U.S. [cited 2023 17 November]. https://www.energy.gov/articles/doe-launches-bipartisan-infrastructure-laws-8-billion-program-clean-hydrogen-hubs-across

Energy Gov (2023) U.S. national clean hydrogen strategy and roadmap

Enhance (2023) Alberta carbon trunk line [cited 2023 13 September]. https://enhanceenergy.com/actl/

ENOC (2023) DEWA and ENOC sign an MoU to develop and operate a joint integrated pilot project for the use of hydrogen in mobility [cited 2023 7 November]. https://www.enoc.com/en/media-centre/news-releases/press-release-detail/id/390/dewa-and-enoc-sign-an-mou-to-develop-and-operate-a-joint-integrated-pilot-project-for-the-use-of-hydrogen-in-mobility

Enterprise Climate (2023) Egypt releases a roadmap to overcoming green hydrogen production obstacles [cited 2023 7 November]. https://climate.enterprise.press/stories/2023/06/14/egypt-releases-a-roadmap-to-overcoming-green-hydrogen-production-obstacles-101590/

equinor (2020) H2H Saltend

Equinor (2023) The Northern Lights project [cited 2023 20 August]. https://www.equinor.com/energy/northern-lights

EUR-Lex (2023) Proposal for a directive of the European Parliament and of the council on common rules for the internal markets in renewable and natural gases and in hydrogen. of application [cited 2023 7 November]. https://eur-lex.europa.eu/legal-content/EN/TXT/?uri=COM%3A2021%3A803%3AFIN&qid=1639664719844

European Commission (2023a) EU taxonomy for sustainable activities [cited 2023 7 November]. https://finance.ec.europa.eu/sustainable-finance/tools-and-standards/eu-taxonomy-sustainable-activities_en

European Commission (2023b) Hydrogen [cited 2023 7 November]. https://energy.ec.europa.eu/topics/energy-systems-integration/hydrogen_en

Fletcher N (2023) Nikola launches hyla brand for hydrogen fuel. 2023-01-26 [cited 2023 17 November]. https://www.ttnews.com/articles/nikola-launches-hyla-brand-hydrogen-fuel

Forbes Middle East (2023) Oman's Hydrom Signs $20B Green Hydrogen Projects Agreements [cited 2023 7 November]. https://www.forbesmiddleeast.com/industry/energy/omans-hydrom-signs-three-agreements-for-green-hydrogen-projects-with-investments-exceeding-$20b

Fuel Cells and Hydrogen 2 Joint Undertaking (2019) Hydrogen roadmap Europe

Global CCS Institute (2022) The U.S. Inflation Reduction Act of 2022 [cited 2023 19 November]. https://www.globalccsinstitute.com/news-media/latest-news/ira2022/

greentechmedia (2020) Shell exploring world's largest green hydrogen project. https://www.greentechmedia.com/articles/read/shell-exploring-worlds-largest-green-hydrogen-project

Gulf News Report (2023) UAE's updated energy strategies to create 50,000 new green jobs by 2030, says Energy Minister [cited 2023 7 November]. https://gulfnews.com/business/energy/uaes-updated-energy-strategies-to-create-50000-new-green-jobs-by-2030-says-energy-minister-1.96777384

H2 View (2020) Preview: Shell exclusive on its latest mega-project, the NortH2 green hydrogen plan. https://www.h2-view.com/story/preview-shell-exclusive-on-its-latest-mega-project-the-north2-green-hydrogen-plan/

H21 (2019) H21 Leeds City Gate

HM Government (2020) The ten point plan for a green industrial revolution

Hydrogen and Fuel Cell Strategy Council (2019) The strategic road map for hydrogen and fuel cells

Hydrogen Central (2022) Germany, Japan tap hydrogen to reduce Russia dependence - hydrogen central. 2022–04–28 [cited 2023 13 September]. https://hydrogen-central.com/germany-japan-hydrogen-russia-dependence/

Hydrogen Central (2022) Saudi Arabia hydrogen industrial strategy. 2022-01-10 [cited 2023 13 September]. https://hydrogen-central.com/saudi-arabia-hydrogen-industrial-strategy/

Hydrogen Energy Industry Promotion Association (2023) Inner Mongolia builds integrated wind and solar hydrogen production demonstration project-Hydrogen Energy Branch of China Industrial Development Promotion Association [cited 2023 23 September]. https://cn-heipa.com/newsinfo/1799691.html

Hydrogeninsight (2022) Fresh blow for hydrogen vehicles as average pump prices in California rise by a third to all-time high. 2022–11–10 [cited 2023 13 September]. https://www.hydrogeninsight.com/transport/exclusive-fresh-blow-for-hydrogen-vehicles-as-average-pump-prices-in-california-rise-by-a-third-to-all-time-high/2-1-1351675

Hydrogeninsight (2023) The number of hydrogen fuel-cell vehicles on the world's roads grew by 40% in 2022, says IEA report. 2023–05–02 [cited 2023 13 September]. https://www.hydrogeninsight.com/transport/the-number-of-hydrogen-fuel-cell-vehicles-on-the-worlds-roads-grew-by-40-in-2022-says-iea-report/2-1-1444069

Hydrogeninsight (2024) ExxonMobil unveils plans to export half a million tonnes of ammonia to Japan from revamped Texas blue hydrogen project | Hydrogen Insight [cited 2024 5 June]. https://www.hydrogeninsight.com/production/exxonmobil-unveils-plans-to-export-half-a-million-tonnes-of-ammonia-to-japan-from-revamped-texas-blue-hydrogen-project/2-1-1617309

HyNet (2018) HyNet North West: from vision to reality

IEA (2008) Section 45Q credit for carbon oxide sequestration. IEA

IEA (2023) Global EV outlook 2023. IEA

Internal Revenue Service (2020) Credit for carbon oxide sequestration

International Council on Clean Transportation (2023) Total cost of ownership of alternative powertrain technologies for Class 8 long-haul trucks in the United States

International Energy Agency (2019) The future of hydrogen

International Energy Agency (2023a) Timely advances in carbon capture, utilisation and storage – The role of CCUS in low-carbon power systems [cited 2023 17 November]. https://www.iea.org/reports/the-role-of-ccus-in-low-carbon-power-systems/timely-advances-in-carbon-capture-utilisation-and-storage

International Energy Agency (2023b) Renewable Hydrogen from Oman

IQAir (2023) India air quality index (AQI) and air pollution information [cited 2023 7 November]. https://www.iqair.com/india

IRENA (2022) Geopolitics of the energy transformation: the hydrogen factor. International Renewable Energy Agency, Abu Dhabi

Kansas Reflector (2023) Thanks to federal tax credits, it's boom time in the Midwest for carbon dioxide pipelines. 2023–07–04 [cited 2023 13 September]. https://kansasreflector.com/2023/07/04/thanks-to-federal-tax-credits-its-boom-time-in-the-midwest-for-carbon-dioxide-pipelines/

Kendall M (2023) Private Communication, W.J. Nuttall, Editor

Kingdom of Saudi Arabia (2023) Saudi Vision 2030 [cited 2023 7 November]. https://www.vision2030.gov.sa/

Kotb M (2023) Egypt considers a potential USD 1 billion green hydrogen plant deal. 2023-08-01 [cited 2023 7 November]. https://egyptianstreets.com/2023/08/01/egypt-considers-a-potential-usd-1-billion-green-hydrogen-plant-deal/

Madhavi M, Nuttall WJ (2019) Coal in the twenty-first century: a climate of change and uncertainty. Proc Inst Civ Eng Energy 172(2):46–63

Major Projects Alberta (2022) Alberta carbon trunk line [cited 2022 20 June]. https://majorprojects.alberta.ca/Details/Alberta-Carbon-Trunk-Line/622

Manthey N (2023) Toyota partners with VDL to roll out heavy-duty hydrogen trucks in Europe I electrive.com [cited 2023 7 November]. https://www.electrive.com/2023/05/10/toyota-partners-with-vdl-to-roll-out-heavy-duty-hydrogen-trucks-in-europe/

Midlands Engine and Green Growth (2021) Hydrogen technology strategy

Ministerial Council on Renewable Energy Hydrogen and Related Issues (2017) Basic hydrogen strategy

Murtaugh D (2023) China approves renewable mega-project for green hydrogen [cited 2023 23 September]. https://www.bqprime.com/china/china-approves-renewable-mega-project-foc used-on-green-hydrogen

National Energy Technology Laboratory (2023) Carbon capture and storage database. https://netl.doe.gov/home

National Petroleum Council (2019) Meeting the dual challenge

National Portal of India (2023) National green hydrogen mission [cited 2023 13 September]. https://www.india.gov.in/spotlight/national-green-hydrogen-mission

Neom (2023) NEOM Green Hydrogen Company completes financial close at a total investment value of USD 8.4 billion in the world's largest carbon-free green hydrogen plant [cited 2023 7 November]. https://www.neom.com/en-us/newsroom/neom-green-hydrogen-investment

News B (2019) China's hydrogen vehicle dream chased with $17 billion of funding, in Bloomberg

Nuttall, W.J., Neppolian, B., Madhavi, M., and Rajan, A. (in press). Prospects for Hydrogen Energy and its Production in India, In Lean, H.H. & Chan J.H. (ed.) Energy Economics, Finance, and Management in Developing and Emerging Countries. Elsevier.

Office of Energy Efficiency & Renewable Energy (2022) Hydrogen pipelines [cited 2022 20 June]. https://www.energy.gov/eere/fuelcells/hydrogen-pipelines

Partidário P et al (2020) The hydrogen roadmap in the Portuguese energy system—developing the P2G case. Int J Hydrogen Energy 45(47):25646–25657

Petrova V (2023) China's Inner Mongolia clears massive green hydrogen plan [cited 2023 13 September]. /news/chinas-inner-mongolia-clears-massive-green-hydrogen-plan-report-751425/

Port of Rotterdam (2020) Port of Rotterdam becomes international hydrogen hub

Robiniusa M et al (2018) Comparative analysis of infrastructures: hydrogen fueling and electric charging of vehicles. Institut für Elektrochemische Verfahrenstechnik

S&P Global (2021a) Gulf of Mexico Lease Sale 257 sees more participation, improved bidding [cited 2022 20 June]. https://www.spglobal.com/commodityinsights/en/market-insights/latest-news/oil/111721-exxonmobil-bids-big-in-texas-shallow-waters-during-us-gulf-lease-sale-257

S&P Global (2021b) US Steel, Equinor team up to examine Appalachia's potential for hydrogen [cited 2023 1 September]. https://www.spglobal.com/marketintelligence/en/news-insights/latest-news-headlines/us-steel-equinor-team-up-to-examine-appalachia-s-potential-for-hyd rogen-65235065

State Information Service (2023) Petroleum minister announces framework of Egypt's low-carbon hydrogen strategy [cited 2023 13 August]. https://www.sis.gov.eg/Story/172771/Petroleum-min ister-announces-framework-of-Egypt%27s-low-carbon-hydrogen-strategy/?lang=en-us

Strategy H (2020) Enabling a low-carbon economy. Energy

Sultinate of Oman (2023) Vision Document

Sun P et al (2019) Criteria air pollutants and greenhouse gas emissions from hydrogen production in U.S. steam methane reforming facilities. Environ Sci Technol 53(12): p. 7103–7113

The Federal Government (2020) The national hydrogen strategy

The Fuel Cell and Hydrogen Energy Association, Road map to a US hydrogen economy

The Hydrogen Council (2020) Path to hydrogen competitiveness: a cost perspective

The Oxford Institute for Energy Studies (2023) Egypt's low carbon hydrogen development prospects [cited 2023 13 August]. https://www.oxfordenergy.org/publications/egypts-low-car bon-hydrogen-development-prospects/

The World Bank (2023) Green growth opportunities in the GCC

U.S. Energy Information Administration (2023) Natural gas wellhead value and marketed production [cited 2023 18 November]. https://www.eia.gov/dnav/ng/ng_prod_whv_a_EPG0_VGM_mmcf_a.htm

United Arab Erimates (2023) National Hydrogen Strategy [cited 2023 7 November]. https://u.ae/en/about-the-uae/strategies-initiatives-and-awards/strategies-plans-and-visions/environment-and-energy/national-hydrogen-strategy

US Department of Energy (2013) Breakthrough large-scale industrial project begins carbon capture and utilization. https://www.energy.gov/fe/articles/breakthrough-large-scale-industrial-project-begins-carbon

US Energy Information Administration (2022) Country analysis executive summary: Australia. US Energy Information Administration

Verheul B (2019) Overview of hydrogen and fuel cell developments in China. Holland Innovation Network China

YouTube (2020) Live: Boris Johnson unveils 'New Deal' style UK infrastructure plan. https://www.youtube.com/watch?v=psFiGjlxlC4

Chapter 5
Innovations to Watch

Abstract This chapter introduces a set of ideas that go beyond those presented in chapter 3. The chapter gives voice to independent experts, sometimes in collaboration with book authors to explore emerging concepts. The ideas presented have the potential to shape the future trajectory of the hydrogen industry. The key concepts include: geological hydrogen, hydrogen and nuclear energy, methane pyrolysis, hydrogen cryomagnetics, and hydrogen for aviation.

In this chapter we showcase some particular aspects of hydrogen innovation. The list is neither exhaustive nor necessarily representative of the most important emerging ideas and technologies. Rather it reflects some of the issues raised in our project webinars and our 2022 event held in London. Where possible, we take the ideas a little further. We suggest that these various innovations are important to consider when looking to the future emergence of a global hydrogen economy. We posit that each of the innovations discussed has the potential to be transformative to some degree.

This book is a monograph authored by those listed in the frontmatter and on the title page. This chapter is something of an exception, however. We seek to hear directly from the experts. While our thoughts, as book authors, are generally in alignment with what the experts told us, we need to be clear what ideas are theirs. To this end we here adopt a text box structure. We use that presentational device to make clear the origins of all advice and opinion presented. In several cases the work represents a collaboration between an external expert and one of the book authors.

Each external contributor to this chapter has kindly approved our publication of their thoughts and ideas and we are most grateful to them for that permission and for their support more generally. They have been true friends to our project. Of course, they have no responsibility for any words or opinions in this book beyond those clearly identified as their own.

Box 5.1: Geological Hydrogen *By Bo Sears and William Nuttall*

One of the enduring myths about hydrogen is that it is an energy carrier like electricity and not a primary fuel. It is frequently asserted that hydrogen for a hydrogen economy needs to be manufactured using another energy source. Such a logic asserts that hydrogen is not a fuel like natural gas that can be extracted from natural geological sources. While it is no doubt true that a large-scale hydrogen economy will require the manufacture of hydrogen at scale, it is an exaggeration to say that commercially viable geological hydrogen does not exist.

Key to understanding the potential for geological hydrogen can be found via the upstream *helium* industry. Historically, helium has been found in commercially viable concentrations in certain natural gas fields. Recent decades have seen significant demand increases in the global helium industry and prices have risen significantly. Upward price pressure has also come from the fact that anticipated sources, like Russia, have been removed from the global market due to the Ukrainian conflict (C&EN 2023). Helium is a gas required for numerous high technology applications including the cooling of the large super-conducting magnets used in medical magnetic resonance imaging (MRI). In recent years demand for helium has been so intense that some end-users have resorted to risking exploration dollars in its pursuit of reliable and adequate supply for future business. These developments in helium (away from being an ancillary consideration on the margins of the natural gas business) anticipated similar developments that came later relating to hydrogen exploration. Although we talk about geological hydrogen as a story from the past, because it is, it must be said that this innovation only started to gain serious traction in the 2020s.

The earliest steps in the story of geological hydrogen were accidental. While drilling for water in Mali in 1987, an operator accidentally discovered an anomalous geological hydrogen resource. Initially, interest in the discovery was limited, but much later, in 2012 serious attention was given and the gas was confirmed as having a composition of 98% pure hydrogen (Ball 2022). To date, this is the purest geological hydrogen resource ever found and it has been used to provide electricity to a local village.

As introduced earlier, the systematic (and commercially motivated) search for geological hydrogen arguably has its roots in the entrepreneurial search for helium. That experience uncovered the disruptive observation that geological hydrogen could exist at scale. The random discovery of geological hydrogen in Mali was not as unusual as might have been thought.

Interestingly the fact that there is geological hydrogen has remained hidden for so long lies in the instrumentation used to test natural gas wells. Generally speaking, a gas chromatograph is used to test the composition of gases in a natural gas field. However, as helium is not necessarily common in most natural gas fields, helium is often used to purge these instruments, thus rendering helium's detection impossible. As hydrogen represents a similar "peak" as helium in a gas chromatograph reading, it is also undetectable using any chromatograph that was cleaned with helium prior to the measurement being made. The point is that if you clean the instrument with helium – you obtain a massive helium peak which the operator is trained to ignore. Under that enormous peak are lost the real readings for helium and hydrogen in the well gas sample. It seems probably that the vast majority of 20th Century oil and natural gas exploration could not see the hydrogen in their test drills. Hence it is quite possible that hydrogen exists in commercial quantities, especially in areas where other "inert" gases are found. In traditional oil and gas-speak, these resources could be classified as "missed pay."

There are extensive records of prospective oil and gas wells from all over the world. Very often they have nothing to say about the light gases. If a full analysis of these gases either via a gas chromatograph or a mass spectrometer were obtained, it is possible that a new asset class of geological hydrogen could be born.

Results from a gas well in the Rocky Mountain region in the United States where care was taken to determine the contribution from light molecules, show that helium is already economic at 2%. This prospect would be a commercial proposition even if it focused solely on helium. However, the presence of hydrogen is over 25%. The hydrogen gas from that Rocky Mountains well is surely a commercial proposition in its own right. Taken together the two light gases make this well very interesting indeed. With helium and hydrogen available in such quantities we see the prospect of a wave of renewed interest in previously neglected opportunities. For those familiar with natural gas wells it is interesting to note that the level of CO_2 is low, at less than 6%.

At the time of writing there is no commercial geological hydrogen production anywhere in the world outside of the small Mali find, but the potential could appear to be enormous, and we expect things to move quickly in this space. In fact, efforts are currently underway to search for geological hydrogen in South Australia and many companies have been formed around this effort. In France government scientists report the discovery of a large geological hydrogen resource under the coal fields of Lorraine in northern France (Bettayeb 2023). The cases of Australia and France take us to the question of why is there geological hydrogen? – More about that later.

It seems that hydrogen exploration is now taking off and in many cases it will be to helium exploration as helium was to natural gas. That is the world will be looking for hydrogen in places that are being considered for helium. The commercial proposition, at least initially, would appear to be co-production primarily with hydrocarbons, but also including co-production with helium driven by its own industry dynamics.

Looking to the helium experience, it is perhaps not too optimistic to expect that North American geological hydrogen resources can be found that are far richer than 25%. These would be exciting commercial prospects – especially if a hydrogen economy starts to emerge at scale in the coming years.

Once one accepts the presence of significant concentrations of helium in geological gases the question soon becomes – what is the precise origin of the hydrogen that we are seeing? The primary method is believed to be Serpentization, which is the hydrothermal alteration of mafic/ultramafic rock causing free hydrogen release. Serpentization is known to occur at the spreading of the ocean floor in the Mid Ocean Ridge. Another method, Radiolysis, is the cracking of water molecules after bombardment from radioactive decay alpha particles. Finally, Shallow Hydrogen is typically found at the top of, or indeed above, natural gas reserves. In this case the hydrogen arises as a consequence of hydrocarbons at deeper strata. Hydrogen rises and hits a higher impermeable layer and is capped. All sources of geological hydrogen are potentially of commercial interest. Some opportunities are closely linked with helium sites while other will be more associated with natural gas. Let us think more deeply about that... The two hydrogen-linked products helium and natural gas could actually be quite distinct propositions.

If our instincts are correct, deep and shallow hydrogen are fundamentally different propositions. The more radical of the two proposals is deep hydrogen. Indeed we suggest that there may be two possible processes giving rise to the deep hydrogen being found co-located with helium.

The first hypothesis for radiogenic deep hydrogen arises from the radiolytic decomposition of water over geological time periods which may have ended millennia ago. For this to occur one would need water to exist in close proximity with naturally occurring radioactive materials (NORM). In contemplating the presence of intense natural radioactivity we note the discovery in the twentieth century of ancient natural nuclear reactors in Oklo, Gabon, West Africa (International Atomic Energy Agency 2018).

The second hypothesis concerning the radiogenic generation of deep hydrogen is the serpentization of ultramafic rocks (essentially the hydrolysis and transformation of ferromagnesian minerals) (McCollom and Bach 2009).

We suggest that commercial exploration for light gases (helium and hydrogen) should be of interest to those seeking to balance risk in their portfolio of energy investments. Light gas exploration and extraction has the potential to hedge key risks efficiently, such as those associated with natural gas derived hydrogen production using steam methane reforming. Such hedging has the potential to be efficient because geological hydrogen production involves no natural gas price risk. Arguably the key risks are actually helium-related and hence largely independent of risks associated with natural gas economics.

In 2023 the mainstream media picked up on the story of the geologically sourced hydrogen. It was soon christened "White Hydrogen" (Guardian

2023) although other commentators have preferred the term "Gold Hydrogen" perhaps intending to suggest that a "Gold Rush" is imminent (Zgonnik 2020; Prinzhofer et al. 2018).

In 2024 excitement around White/Gold Hydrogen grew strongly. Norwegian headquartered energy intelligence company Rystad Energy started to talk explicitly about a new white hydrogen gold rush – describing the emergence of this new resource as a 'gamechanger' (RystadEnergy 2024). In part global enthusiasm has been driven by major successes in dedicated exploration efforts. In May 2024 Australian company Gold Hydrogen Limited announced the discovery at its Ramsey 2 facility on the Yorke Peninsula in South Australia, of rich co-located deposits of helium and hydrogen (Gold Hydrogen 2024). The company reports hydrogen available at very high purity (up to 95.8%). Meanwhile back in the US interest continues to grow.

Box 5.2: Hydrogen and Nuclear Energy *By William Nuttall*

In Chapter 1 we introduced the notion of Red, Purple and Pink hydrogen as relating to hydrogen from nuclear energy. Such a brief presentation, however, fails to capture the full range of opportunities in this space.

Of course, one can use nuclear generated electricity from conventional nuclear power stations to electrolyse water for hydrogen. Indeed in April 2022 French nuclear energy conglomerate the EDF Group announced an international plan to develop 3GW of hydrogen capacity by 2030 (EDF 2022). In the UK EDF Energy plans to implement such technology at the proposed new nuclear power plant, Sizewell C, following demonstration of small-scale capacity (<10 MW) at the adjacent and currently operational Sizewell B plant (Sizewell 2023). The logic behind EDFs plans is that nuclear power stations have an economic incentive to operate at a constant rate of energy conversion. Indeed, the plants are also technically at their best when in this condition. This is what has prompted nuclear power traditionally to play the role of baseload generator in the power system. In the future, with the growth of varying renewables on the supply side and with continuing power demand variability there will be time when national power production far exceeds instantaneous demand. At such times it makes sense to divert power away from the grid and into storage and hydrogen production. What is not so clear is why such electrolyser-based hydrogen production need be undertaken at the nuclear power station site or even by the nuclear electricity business. One might imagine an entirely merchant electrolyser operation at a remote location working solely on the basis of market price signals. The nuclear power plant site does have the benefit of abundant low-grade heat, but it is not obvious how helpful such aspects actually are. Hydrogen generated by electrolysis using nuclear power is sometimes termed 'pink hydrogen', as discussed in Chapter 1.

A more radical approach to nuclear hydrogen production has long been advocated by the American company General Atomics. That approach is based upon the use of high temperature nuclear reactors (Nuttall 2022). Rather than separate water into hydrogen and oxygen via electrolysis, if sufficiently high temperatures are available, then it becomes possible to split water directly through a series of high-temperature chemical reactions. This is the thermo-chemical splitting of water and is shown in Fig. 5.1 (taken from Onuki et al. (2009)). While some such approaches do still include a role for electricity others do not. For example, the sulphur-iodine cycle long-considered by General Atomics requires no electricity (nuclear or otherwise), but it does requires an industrial process heat source of at least 850°C. Key attributes of the proposed S-I cycle are described in a conference presentation by Paul Matthias and Lloyd Brown from 2003 (Mathias and Brown 2003).

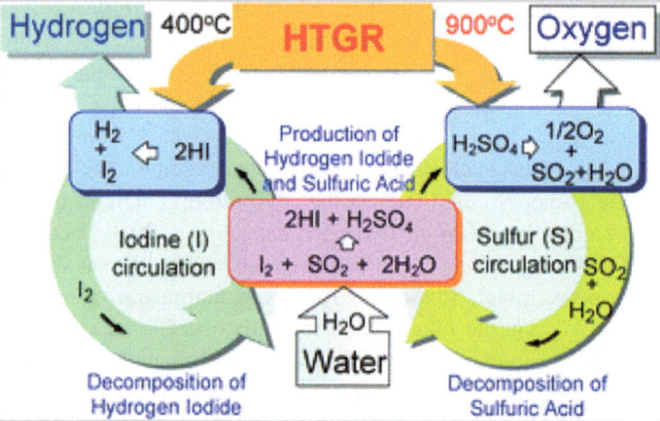

Fig. 5.1 Schematic representation of water splitting via the sulphur-iodine cycle, as driven by a High Temperature Gas-Cooled Reactor (HTGR) Source: (Onuki et al. 2009)

While much High Temperature Gas-Cooled Reactor (HTGR) research and development is underway in Japan, China and now to a lesser extent in the UK (where there are synergies with the UK's long-standing experience with the Advanced Gas-cooled Reactor (AGR)), there has, as yet, been no industrial scale demonstration of the sulphur-iodine cycle driven by nuclear process heat. The proposition remains very much a technology for the future.

Hydrogen and Nuclear Fusion by William Nuttall

Looking to the future there is the possibility of the commercial production of nuclear fusion energy. The orthodox view has been that this new nuclear energy source will be used to generate low carbon electricity. With colleagues, I have long taken the view that the more likely initial large-scale commercialization of fusion energy will be via process heat applications, especially hydrogen production (Nuttall, et al. 2020).

Low carbon hydrogen might be produced using fusion energy either via thermochemical water splitting or through boosted methane reforming with carbon capture and storage. The main challenges associated with the case for fusion as an electricity technology are the need to achieve high levels of reliability sufficient to enable a baseload generation business model like that used by nuclear-fission-based power plants today. For fusion, I suggest that such reliability will be a daunting challenge. Second fusion power technology will arrive only after the electricity system has already been decarbonised and competing options (e.g. renewables and fission based nuclear power) have had a chance to drive down the cost of low-carbon electricity significantly. Notwithstanding the technological immaturity of fusion technology compared to the fission alternative, some potential benefits do exist for fusion over fission as an energy source, and most especially for hydrogen production. These are primarily in the area of system safety. The safety case for a hybrid fusion-chemical plant can be expected to be easier than for a hybrid fission-chemical plant. A fusion reactor involves enormously less stored energy in the reactor core and the radioactive material inventory associated with the reactor is also vastly less than in the case of fission. A fission reactor needs care and attention even when in a shutdown state. In particular, for today's fission reactors post-shut down power must be supplied to drive cooling systems sufficient to manage decay-heat from the nuclear fuel. If the reactor is not actively cooled, decay heat is sufficient to melt the reactor core of a fission power plant as occurred at Three Mile Island in Pennsylvania in 1979 and at the Fukushima-Daiichi plant in Japan in 2011. The use of fusion reactor technology would avoid these risks entirely. Once shutdown a fusion reactor needs no active care. Finally, a fusion reactor involves no actinide materials, such as uranium and plutonium. This also simplifies safeguards arrangements associated with nuclear weapons proliferation prevention compared to the fission reactor alternative, but the technology of nuclear fusion should not be regarded as being entirely free of proliferation concern given that in its most researched form (deuterium–tritium fusion) it has the potential to be a significant neutron source and hence measures should be adopted to ensure that no covert fissile material (usable in nuclear weapons) is produced using fusion facilities.

When considering the future prospects for nuclear fusion energy it is important to reflect on the claimed credentials of the new technology. It is often reported that the primary fuels for the most well-developed fusion energy approach, namely deuterium (a form of hydrogen) and lithium (as currently widely used in battery technology and from which tritium, the proximate fuel for fusion, is created) are both abundantly available and affordable. The statement is true and indeed it favours fusion as a long-term energy option. What has been less discussed is the availability of liquid helium to cool the high-power superconducting magnets that are required by the most highly developed fusion energy concept – the tokamak among others. In the orthodox approach liquid helium is the key consumable for tokamak fusion. The problem is that helium is a by-product of today's fossil fuel economy (natural gas extraction). Prices of helium have been volatile in recent decades and there is much uncertainty around future supply (see text Box 5.1). Such ideas led Bartek Glowacki to propose that cryogenic liquid hydrogen should be used in place of liquid helium to cool the magnet coils – "hydrogen cryomagnetics", see box 5.4). For a fusion based facility producing liquid hydrogen commercially for sale access to this alternative cryogen would not be a problem. The idea of a fusion tokamak producing and using liquid hydrogen was introduced by Nuttall, Glowacki and Clarke in 2005 (Nuttall 2005). We coined the term "Fusion Island" for our ideas.

I see the potential for an important linkage between the future of fusion energy and the future of the hydrogen economy. Fusion will struggle to undercut the relatively low price of Blue Hydrogen, but the possibility of synergistic innovation between fusion and what we might call "second generation Blue Hydrogen" (involving the use of low carbon process heat support) could prove to be particularly advantageous. As noted above, fusion is a much easier technology to integrate with a chemical engineering plan than a conventional nuclear fission reactor, for regulatory reasons alone.

Box 5.3: Hydrogen and Solid Carbon *Karim Anaya and William Nuttall*

As introduced in Chapter 1 the world of hydrogen supply has been systematized according to a range of colours associated to the ways in which the gas is produced (i.e. process and source). We have Grey or Brown (currently the dominant and cheaper highly polluting approach) and Pink, Green or Blue (generally the cleaner, more expensive and more innovative technologies). We have so far considered several of these production routes in this book, but a different way to produce clean hydrogen is via methane pyrolysis, now known as Turquoise Hydrogen. To recap, the Table 5.1 summarises the sources, processes and products related to each type of hydrogen production.

Table 5.1 Main characteristics of hydrogen production [Author's own work [KA]]

Type of hydrogen	Source					Process					Product			
	methane	coal	bio methane	nuclear energy	renewable energy	SMR	gasification	SMR with CCUS	electrolysis	pyrolysis	H_2	CO_2	O_2	solid carbon
Grey	X					X					X	X		
Brown		X					X				X	X		
Blue	X	X (or)						X			X			
Pink				X					X		X		X	
Green					X				X		X		X	
Turquoise	X		X (or)							X	X			X

The production of Turquoise Hydrogen requires a thermal process called methane pyrolysis where methane (the main natural gas component) is split into hydrogen and solid carbon. There are different pyrolysis methods to crack methane, among them are molten bath (with tin), catalytic and plasma (microwave, DC, AC). Plasma is among the most promising. Plasma is referred as the "fourth state of matter" and is created when any gas (i.e. methane in this case) is heated at high temperatures. Plasma is composed of positive (ions) and negative (electrons). The created gas is filtered, and solid carbon is separated and stored while hydrogen can be stored or processed to increase its concentration (purification system). Depending on its uses, different levels of purification can be required (i.e. hydrogen for mobility).

Carbon intensity (i.e. kg CO_2e/kg H_2) is much lower with pyrolysis (0.45) than with SMR only (11) and SMR + CCUS (3), and negative (-2) if renewable natural gas (i.e. biomethane) is used as feedstock. It also requires less use of water and electricity than electrolysis (approx. one seventh) (Monolith 2023).

There are different Turquoise Hydrogen production initiatives and projects around the world. Monolith, a US based turquoise hydrogen producer firm, has received conditional approval for a US\$1.04 billion loan from the US Department of Energy's Title XVII Innovative Energy Loan Guarantee Program to expand the production facilities of clean hydrogen and carbon black in Nebraska (Monolith 2021). The pyrolysis technology developed by Monolith uses 100% renewable energy to convert natural gas into hydrogen and carbon black. After the expansion completion in 2025, most of hydrogen will be used to produce clean ammonia (c. 275,000t pa) while carbon black (c. 194,000t pa) will be sold to leading tire manufacturers such as Goodyear and Michelin (Ammonia Energy Association 2022).

In France, Plenesys is developing a methane pyrolysis process using AC plasma torch. The first commercial scale units (smallest units) need around 29 kg/hr of methane flow rate to produce 7 kg/hr H2 and have 100 kW of plasma torch nominal power. In the USA, the MIT's Carbon Capture, Utilization and Storage Centre is working on a research project to produce hydrogen with no CO_2 emissions using methane pyrolysis with inert molten tin instead. This method helps to prevent the formation of solid by-product on the reactor walls. The Centre has been awarded US\$ 0.75 million for the project (Payne 2021).

Box 5.4 Hydrogen Cryomagnetics *By Bartek Glowacki and William Nuttall*

Over many years we have stressed a potentially important consideration that could greatly enhance the potential for the hydrogen economy. We observe the potential for cryogenic liquid hydrogen at a temperature of 20.4 K at atmospheric pressure, or if pumped to a lower pressure at around 15 K, to cool magnets based on medium and high temperature superconductivity. Such high field magnets can, for instance, be extremely well suited to high-torque electric motors, magnetic storage systems, fusion magnets and magnetic separators. We call such a use for hydrogen 'hydrogen cryomagnetics' (Nuttall and Glowacki 2010). The hydrogen acts simultaneously as a coolant (e.g., for electromagnets, such as those found in electric motors) and as a fuel in combination with oxidant to power the system or device. With the best High T_c superconducting materials, we expect that it is possible to achieve fields as strong as 40 Tesla via hydrogen cryomagnetics.

We suggest a range of possible applications for hydrogen cryomagnetics:

- Fully superconducting electric motors for regional and short-range aviation – see text box 5.5.
- Offshore wind turbines for local desalted seawater electrolysis with local hydrogen liquefaction. Hydrogen cryomagnetics can easily make possible 10 MW wind turbine technology, and possibly go as far as 25 or 30 MW. This is because the superconducting generator is much more compact and less heavy. The world should aim to stop making, and deploying, conventional 5MWe wind turbines, there is so much lifecycle CO_2 associated with making them, that it would surely be better to make a 25 MWe turbine at the outset.
- City and long-distance electricity feeds. Currently undergrounded city electricity feeds are based on oil-cooled copper wire. In the future liquid hydrogen could be used to cool a high voltage DC transmission feed cable based on the use of high temperature superconducting material. While this proposition is not strictly an example of hydrogen cryomagnetics, there are key synergies to be exploited in such an endeavour. This infrastructure would supply both electricity and hydrogen energy to the city from the place of electricity and hydrogen generation. Japan has much research and development oriented to such ideas. The potential for high temperature superconductivity in future power cables has been reviewed by Mohammad Yazdani-Asrami and co-workers in 2022 (Yazdani-Asrami et al. 2022).
- The non-ferrous metals industry could deploy hydrogen cryomagnetics to improve superconducting induction melting. A magnetic non-ferrous ignot can be rotated in a very strong magnetic field as might be supplied using hydrogen cryomagnetics. The metal billet can then be spun at high rotational speed in the strong magnetic field and rapid heating results sufficient to permit metal work or even to melt the billet. The process described above only achieves competitive viability if fields of at least 3 T can be generated.

This is essentially impossible for conventional electromagnets at scale, but it would be straightforward for hydrogen cryomagnetic systems.

- Superconducting mechanical energy storage based on low friction flywheels and superconducting bearings and similar technologies would be greatly enabled by hydrogen cryomagnetics.
- Hydrogen cryomagnetics has also been suggested as a key enabler of a new distributed hydrogen economy based upon the local production and use of liquid hydrogen (Glowacki et al. 2015).
- Finally the application which first prompted us to consider hydrogen cryomagnetics was a tokamak fusion facility developed so as to manufacture liquid hydrogen as a commercial product and to use the same product as the cryogen for the large tokamak magnets. We call this concept – "Fusion Island" (Nuttall 2005). Central to the hydrogen cryomagnetics proposition is the commercial production of cryogenic liquid hydrogen (LH_2) for distribution in a manner similar to cryogenic liquefied natural gas today.

Traditionally LH_2 production only works at scale, even at a production rate of 10,000 kg/hr one loses 30% of one's initial energy simply via the act of liquefying (note: 1 kg = 14 L of LH_2). This is because of the need to navigate a nuclear magnetic reality known as the ortho-para transition (Nuttall et al. 2016).

Figure 5.2 points to the possibility of, and potential for, a new "advanced" thermo-magnetic method suitable for small scale hydrogen liquefaction at high efficiency. The emergence of thermodynamically and economically viable local liquid hydrogen production at the scale of roughly 5 L/hr could hugely boost the prospects for decentralised hydrogen cryomagnetic innovation, roll-out and application.

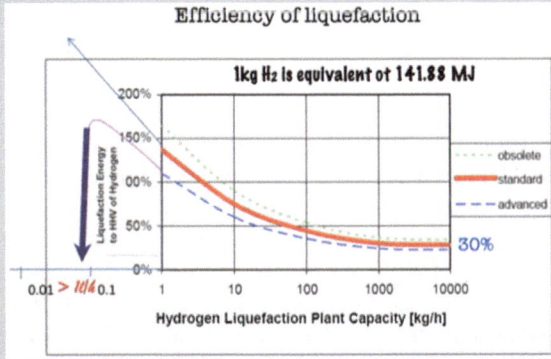

Fig. 5.2 Projected Liquefaction Energy for decentralised economy at 0.1 kg of hydrogen per hour, based on novel advanced technologies prompted by scientific developments by Prof. B.A. Glowacki, University of Cambridge. Image Source: BA Glowacki

Back in the late 1980s, when high temperature superconductors were first discovered, the conventional wisdom was that liquid nitrogen cryogenics would be key to their successful deployment. We suggest that this was incorrect. Liquid hydrogen at 20K will be the key cryogen for high temperature super-conductivity in the twenty-first century. The reason being it frees us of the problems of liquid helium availability and price and yet also provides cooling to temperatures far lower than T_c. Such a temperature regime permits very high current densities and very high magnetic field generation. In most cases this is not possible with liquid nitrogen at 77 K.

Our academic comments concerning the future potential of hydrogen cryo-magnetics must be balanced by industrial realities. Is the emergence of a liquid hydrogen economy a credible proposition? There are already initial indications that it might be. We note with great interest the recent development in Japan of a liquid hydrogen cargo ship – the Suiso Frontier. This demonstration of technology could be a sign of things to come (Čučuk 2023).

Separately, Shell has, with partners, been developing a liquid hydrogen cargo containment system (DNV 2024). Working with technical experts CB&I, Shell is keen to learn from that company's experience with static, land-based, liquid hydrogen storage when designing for commercial LH2 cargoes in a maritime context. A third partner, DNV, is bringing in its regulatory knowledge and experience to help ensure a smooth way ahead for the technical innovations involved.

Box 5.5: Hydrogen and Aviation *By William Nuttall and Bartek Glowacki*

In Chapter 3 we introduced aviation as a key transport sector open to hydrogen-based innovation. In this box we will explain the important potential for hydrogen cryomagnetics in low-carbon aviation. First, however, we should set the scene.

Aviation is widely regarded as one of the most difficult to decarbonise sectors of our modern economy. The gas turbine jet engine fuelled with kerosene has transformed the human experience over the last 70 years. Cargo and people can reach any part of the world in a matter of hours.

The leading approach to reducing aviation emissions is the use of sustainable biofuel derived from plant-based oils or from solid biomass via pyrolysis. Typically, such fuels are used as a blend with conventional JET-A unleaded kerosene-based aviation fuel. Modern airliners are authorised to operate with up to 50% sustainable biofuel in the blend (ASTM 2022).

The goal has been to find a sustainable JET-A alternative that can be used in modern aircraft with minimal adjustments required. A second candidate approach has been the proposed development of electrofuels, or e-fuels, as discussed in Chapter 2. That is the use of captured carbon together with hydrogen to produce a synthetic hydrocarbon fuel with minimal environmental impact. Noting the complexity of such a process and the relatively low cost of conventional JET-A fuel such measures appear to be far from economically attractive. Thus far little progress has been made, for aviation at least.

Thus far we have considered ways to fuel today's gas turbine jet engines. A more ambitious approach is to abandon the use of kerosene (natural, bio-derived or synthetic) entirely and to consider the use of a zero-carbon fuel directly. Such a fuel could be hydrogen. As introduced in Chapter 3, Hydrogen can be used directly in two ways to propel an aircraft:

Hydrogen fuel cells converting energy stored in hydrogen into electrical energy to power electric motors.

In September 2023 the first successful test flights of a liquid hydrogen fuelled aircraft has been performed in Slovakia (H2FLY 2023). The HY4 demonstrator aircraft uses a fuel cell electric power system and has been developed by a German consortium funded by the German government.

Hydrogen combustion via a similar process to conventional internal combustion (gas turbine or piston engine), this technology generates thrust by burning hydrogen in modified aircraft engines;

The reduction in global greenhouse gas emissions is, of course, the driving motivation for such shifts, but some care is needed. Water vapour emitted from hydrogen powered aircraft can act as a greenhouse gas depending strongly with flight altitude (Nuttall and Glowacki 2010), noting that aircraft contrails from the combustion of kerosene based aviation fuel is already a topic of significant climate concern (International Air Transport Association 2024).

In the context of combustion, around the world major engineering firms such as GE, Mitsubishi and Siemens are working hard on hydrogen fuelled gas turbines (GE Gas Power 2023; Nature Portfolio 2023; Siemens Energy 2023). The main activity has been for stationary applications (e.g. power generation and linking to hydrogen-based energy storage) but around the world aviation applications are definitely under consideration, see e.g. (Rolls Royce 2022).

In September 2023 Rolls-Royce announced that they had made a major step forward in 100% hydrogen combustion for aircraft propulsion (Loughran 2023). The test of a new spray nozzle configuration in a modified Pearl 700 engine showed that hydrogen fuelled combustion can achieve conditions of maximum take-off thrust. The technical challenge relates to the higher temperatures and faster rate of burn seen with hydrogen compared with conventional JET-A (kerosene) fuel.

Over the last several decades, aircraft engines have become larger as a consequence of ever more sophisticated fan blade systems. These fan blades are powered by the gas turbine at the heart of the engine, but it is not the jet action that gives the plane most of its push directly. Rather the fan is acting in a manner not unlike the propellors of the early twentieth century. One potential for hydrogen is to drive the blades of a fan or a propellor directly without the use of a gas turbine at all. One way to do that would be with a high-torque superconducting electric motor. That motor might be energised using electricity from an on-board hydrogen fuel cell (or perhaps from a dedicated hydrogen gas turbine elsewhere in the aircraft). Rather than engineer a dedicated liquid helium cooling system for the high torque motors, real advantages can be obtained by moving to a hydrogen cryomagnetics approach (see Box 5.4). It is the prospect of hydrogen cryomagnetics that represent, in our opinion, the most promising opportunities for low-carbon innovation in aviation.

Aviation – the "killer app" for Hydrogen Cryomagnetics.

A key innovation is emerging - the use in aircraft of hydrogen as a coolant for fully superconducting motors where liquid or slush hydrogen is used to provide cryogenic temperature around 20 Kelvin. The outstanding efficiency of superconducting MW-range motors for distributed power architecture on board electric aeroplanes has been proven already. It has the potential to be a key technology in the future medium range low-emission aircraft. Whatever hydrogen evaporates during motor cooling cycle can be used to power the electricity generation for the aircraft need via fuel cells, a centralised hydrogen gas turbine, or both. In Europe the large Industrial-University consortium managed to succeed in building novel fully superconducting 1 MW motor cooled by on board liquid hydrogen system described in detail in Grilli et al. (2020) (Fig. 5.3).

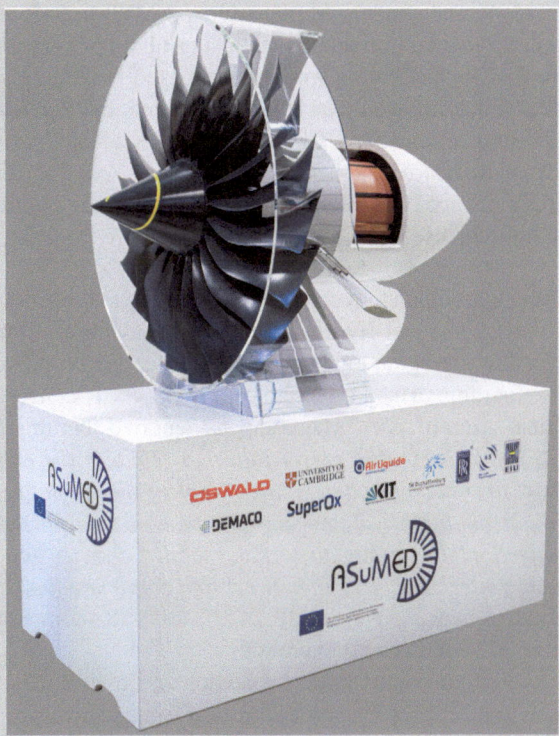

Fig. 5.3 Mock-up of a fully superconducting liquid hydrogen cooled 1 MW motor presented at the 2019 Hannover Fair. Source (Ball 2022), Image rights: IOP Publishing via STM Permissions Guidelines

As noted in Chapter 3, one challenge associated with the use of cryogenic technologies in aircraft is the need to ensure good metallurgical properties for the airframe and flying surfaces. There is a need to preserve the flexibility and strength of these structural materials and to avoid any brittleness that can occur if these materials drop to very low temperatures.

Industrial interest in hydrogen fuelled aero-engines and in liquid hydrogen as an on-board fuel are growing quickly. See for example these reports from the summer of 2023: (Green Car Congress 2023; Malayil 2023). Writing in Aviation Week in April 2001, Thierry Dubois reported (Dubois 2021):

> *"Airbus has launched an ambitious demonstration program for the use of superconducting technology. It is aiming at a major efficiency improvement. The idea stems from both the difficulty of designing an electric-propulsion architecture with conventional wiring and the opportunity to use liquid hydrogen as a cold source. Superconducting materials require cryogenic temperatures."*

In May 2024 Airbus announced plans for a 2 MW aircraft engine based on hydrogen cryomagnetic principles and using high temperature superconducting tapes. Proximate cooling would be achieved by helium recirculation but cryogenic engineering would be based upon liquid hydrogen. Airbus calls the proposed system architecture 'Cryoprop' (FlightGlobal 2024).

Meanwhile in Russia researchers at the Moscow Aviation Institute have started to design a 5 MW fully superconducting liquid hydrogen cooled generator suitable for use as an aircraft propellor drive (Dezhin and Ilyasov 2022).

One company that sought to accelerate the transition to low carbon hydrogen fuelled flight was Universal Hydrogen based in Hawthorne California. The company was developing a modularised approach to on-board hydrogen storage and refuelling. The first product was to be a conversion kit for existing regional aircraft types. The company reported that this proposition: *"consists of a fuel cell electric powertrain that replaces the existing turboprop engines. It also accommodates, in the rear of the fuselage, our proprietary, lightweight, modular hydrogen capsules that are transported from green hydrogen production sites to the airport and loaded directly into the aircraft using the existing intermodal freight network and cargo handling equipment."* (Universal Hydrogen 2023; Team-BHP 2023). Sadly the company ran out of funding in June 2024 and ceased operations.

With thanks to John-Paul Clarke for advice and assistance. The authors (WJN and BAG) take all responsibility for what is presented here.

Box 5.6: Hydrogen and Internal Combustion Innovation *By William Nuttall*

The entire project reported in this volume has grappled with a key question the extent to which the future of the global energy system will be centred upon electrification. Hydrogen is presented as an alternative, but is hydrogen itself merely part of an electrification paradigm involving renewable energy and fuel cell applications or is hydrogen the means to continue the global natural gas industry in an environmentally responsible way, through the use of carbon capture and storage? Blue Hydrogen has the feel of being the more incremental and less radically disruptive way to produce hydrogen in the future. Concerning hydrogen for mobility is there a parallel to be seen as regards hydrogen use? It would appear that there is.

Amidst all the discussion around the potential for hydrogen fuel cells to propel the electric vehicles of the future; one can forget that hydrogen has the potential to decarbonize much more conventional internal combustion engines (ICEs). In this text box we shall focus on two types of ICE in particular- piston engines (PEs) as found in cars and trucks and gas turbines (GTs) as found in aircraft and some ships.

As regards the competition between FCEV and ICE-PE in cars, it appears that, in Europe at least, the FCEV option is favoured by the policy emphasis currently placed on Battery Electric Vehicle Solutions. That is grounded in the view that the future will be based on near-total electrification of the energy economy. If, however, hydrogen emerges as a major clean energy carrier then, if it does not come too late, one might see the emergence of hydrogen-fuelled ICE-PE power trains.

The first major motor manufacturer to explore hydrogen ICE-PE approaches was BMW with its Hydrogen-7 demonstration vehicle produced between 2005 and 2007. These vehicles had the potential to be fuelled by either hydrogen or gasoline. The vehicles had two fuel tanks and the ignition timings were adjustable to suit both fuels. The ability to run on gasoline, if hydrogen is not available, greatly reduces issues of "range anxiety" for motorists planning long-journeys. The Hydrogen-7 project did not get past the initial plan to produce 100 demonstration vehicles. BMW abandoned the ICE approach in favour of FCEV technology. It is reported that BMW turned away from the ICE approach following a critical assessment by the US Environmental Protection Agency which found that even in hydrogen mode the use of oil-based lubricants (with some inevitable combustion of those molecules) meant that the vehicle could not be said to have zero CO_2 tailpipe emissions (Nica 2016).

Hydrogen also presents particular difficulties concerning combustion control and engine management and these can present as reliability challenges. The difficulties of engine management are described in a paper by Wolfram Enke and colleagues (Enke et al. 2007).

While the use of hydrogen as a fuel in an ICE inevitably means that the tailpipe is carbon free (neglecting the very small lubrication-related contaminations discussed above) one cannot be entirely relaxed about the tailpipe emissions. Care has to be taken in engine design and operation that harmful nitrous oxides are not generate by high engine temperatures. The nitrogen is found in the air used for the combustion.

In the United States large specialist power technology company Cummins Inc. has been developing a new hydrogen fuelled ICE-PE. The company reports success with a new design achieving near zero CO_2 emissions through the tailpipe and near zero levels of NOx (Cummins Inc 2021). In 2022 products include the Cummins B6.7H hydrogen engine for medium-duty trucks and the larger Cummins' X15H hydrogen engine aimed at heavy-duty trucks up to 44 T GVW, with a top rating of 530 hp (395 kW) and a peak torque of 2600 Nm (Cummins Inc 2023a). In September 2022, Alison Trueblood, Cummins Executive Director – On-Highway Business Europe said: "*a hydrogen combustion engine fits in today's vehicles, works with today's transmissions, and integrates seamlessly into the industry's existing service networks and practices*".

In the same month Werner Enterprises, a major US logistics provider, signed up to buy 500 units of the larger X15H hydrogen fuelled ICE-PE technology (Cummins Inc 2023b). As we look at the future of heavy transport, especially in the US, we have already said elsewhere that FCEV technology is likely to beat BEV technology. The interesting thing will be whether hydrogen fuelled ICE technology can beat FCEV technology.

Meanwhile in Japan, in November 2021, car manufacturers Toyota, Subaru, and Mazda together with engine manufacturers Kawasaki and Yamaha announced an intention to work together on environmentally benign ways to keep internal combustion technology going (Bigg 2021). One key low-carbon combustion fuel would be hydrogen. In that spirit Yamaha, in partnership with Toyota, has developed a 450HP hydrogen V8 engine. Toyota is prominent among global car companies in not seeing the future as being inevitably Battery Electric led. At the risk of digressing, it is worth pointing out that another low carbon combustion fuel for ICE-PE systems is M-85 fuel a blend of 85% methanol with gasoline. This fuel has the benefit that it is liquid at room temperature greatly simplifying fuel storage and range challenges, albeit at the price that some significant CO_2 emissions are released, albeit far less than from conventionally fuelled ICE systems.

Generally, it is reported that hydrogen works best for longest distance and the heaviest transport requirements. Indeed, this book has said as much, but we must concede that such assessments may not survive in the face of hydrogen-ICE innovation. In April 2021 Israeli engineering firm Aquarius Engines revealed its new hydrogen fuelled 10 kg single-piston-linear-engine. The engine is extremely simple with only 20 components and only one moving part. The company points out that its technology is far cheaper to construct and operate than low carbon FCEV alternatives (Young 2021).

Superturbo Technologies of Loveland Colorado is developing a "*mechanically driven turbocharger system offering bidirectional power transfer and speed ratio control [that] can improve air management in H2 engines*" (SuperTurbo Technologies 2023). This is something that the company calls Hydrogen Boosting and it offers the prospect of easily reducing problematic NO_x emissions from next generation hydrogen ICE-PE systems.

Another emerging area of interest is the potential development of hydrogen fueled opposed piston ICE powertrains. Opposed piston engines typically compress and combust a petroleum fuel (typically diesel) between two symmetrically converging pistons. A conventional ICE system uses a single piston in a given cylinder and fuel air compression ignition occurs between a moving piston and a cylinder head. Opposed piston designs have long been used in large scale ICE systems, such as stationary or marine applications. In February 2024 Argonne National Laboratory (part of the US Department of Energy) and Archates Power revealed plans to develop a two-stroke hydrogen-fuelled opposed piston ICE for heavy duty applications, such as large trucks (Green

Car Congress 2024). It is hoped that the team will be able to benefit from the heat efficiencies that are usually seen in opposed piston engine designs.

As is being seen in the other technology areas discussed in this Chapter we find that industrial innovation around hydrogen internal combustion engines is growing very quickly in the summer of 2023. We note for example these reports: (News 2023; Adler 2022).

We also recall the innovation noted in Chapter 3 when considering Speciality Vehicles under development by the construction equipment sector, for example: (The Construction Index 2023).

In this chapter we have considered a range of over the horizon innovation areas for the emergent hydrogen economy. In Chapter 6 we will reflect back on the full range of issues considered in this book and offer our best *insights into the new hydrogen economy.*

References

ACT News (2023) Hydrogen internal combustion mobility is here. [cited 2023 14 August]. https://www.act-news.com/news/hydrogen-internal-combustion-mobility-is-here/.

Adler A (2022) Werner signs for 500 Cummins hydrogen-powered internal combustion engines. 2022–09–07 [cited 2023 14 August]. https://www.freightwaves.com/news/werner-signs-for-500-cummins-hydrogen-powered-internal-combustion-engines.

Ammonia Energy Association (2022) Monolith Materials: new deal with Goodyear, $1 billion loan from DoE â€" Ammonia Energy Association. [cited 2023 23 August]. https://www.ammoniaenergy.org/articles/monolith-materials-new-deal-with-goodyear-1-billion-loan-from-doe/.

ASTM (2022) Standard Specification for Aviation Turbine Fuel Containing Synthesized Hydrocarbons, in D7566-22

Ball PJ, KC (2022) Natural hydrogen: the new frontier - GEOSCIENTIST. 2022–03–01 [cited 2023 3 September]. https://geoscientist.online/sections/unearthed/natural-hydrogen-the-new-frontier/.

Bettayeb K (2023) A gigantic hydrogen deposit in northeast France? [cited 2023 18 November]. https://news.cnrs.fr/articles/a-gigantic-hydrogen-deposit-in-northeast-france.

Bigg M (2021) Toyota, Subaru, And Mazda are on a mission to save the combustion engine. 2021–11–16 [cited 2023 14 August]. https://carbuzz.com/news/toyota-subaru-and-mazda-are-on-a-mission-to-save-the-combustion-engine.

C&EN (2023) War in Ukraine makes helium shortage more dire [cited 2023 3 September]. https://cen.acs.org/business/specialty-chemicals/War-Ukraine-makes-helium-shortage-more-dire/100/i10.

Čučuk A (2023) Oman welcomes world's first liquefied hydrogen vessel Suiso Frontier. 2023-08-17 [cited 2023 23 August]. https://www.offshore-energy.biz/oman-welcomes-worlds-first-liquefied-hydrogen-vessel-suiso-frontier/.

Cummins Inc. (2021) Cummins begins testing of hydrogen fueled internal combustion engine [cited 2023 14 August]. https://www.cummins.com/news/releases/2021/07/13/cummins-begins-testing-hydrogen-fueled-internal-combustion-engine.

Cummins Inc. (2023a) Cummins Fuels Hydrogen Commitment at IAA. [cited 2023 14 August]. https://www.cummins.com/news/releases/2022/09/20/cummins-fuels-hydrogen-commitment-iaa.

Cummins Inc. (2023b) Werner Enterprises signs letter of intent planning to secure 500 X15H engines from Cummins. [cited 2023 14 August]. https://www.cummins.com/news/releases/2022/09/07/werner-enterprises-signs-letter-intent-planning-secure-500-x15h-engines.

Dezhin, D. and R. Ilyasov (2022) Development of fully superconducting 5 MW aviation generator with liquid hydrogen cooling. EUREKA: Physics and Engineering, 2022(1): p. 62–73.

DNV (2024) Paving the way for large-scale transportation of liquid hydrogen. [cited 2024 5 June]. https://www.dnv.com/expert-story/maritime-impact/paving-the-way-for-large-scale-transportation-of-liquid-hydrogen/.

Dubois T (2021) Airbus' hydrogen drive will materialize in demonstrators. [cited 2023 23 August]. https://aviationweek.com/special-topics/sustainability/airbus-hydrogen-drive-will-materialize-demonstrators

EDF (2022) The EDF Group launches a new industrial plan to produce 100% low-carbon hydrogen. 2022–04–13 [cited 2023 23 August]. https://www.edf.fr/en/the-edf-group/dedicated-sections/journalists/all-press-releases/the-edf-group-launches-a-new-industrial-plan-to-produce-100-low-carbon-hydrogen.

Enke W et al (2007) The Bi-fuel V12 engine of the new BMW Hydrogen 7. MTZ Worldwide 68(6):6–9

FlightGlobal (2024) Airbus developing 2MW superconducting powertrain demonstrator for hydrogen aircraft. [cited 2024 8 June]. https://www.flightglobal.com/aerospace/airbus-developing-2mw-superconducting-powertrain-demonstrator-for-hydrogen-aircraft/158418.article.

GE Gas Power (2023) Hydrogen fueled gas turbines. [cited 2023 23 August]. https://www.ge.com/gas-power/future-of-energy/hydrogen-fueled-gas-turbines

Glowacki BA et al (2015) Hydrogen Cryomagnetics for Decentralised Energy Management and Superconductivity. J Supercond Novel Magn 28(2):561–571

Gold Hydrogen (2024) Hydrogen purity hits new highs in South Australia testing. [cited 2024 5 June]. https://www.goldhydrogen.com.au/updates/hydrogen-purity-hits-new-highs-in-south-australia-testing/.

Green Car Congress (2023) Rolls-Royce awarded £82.8M for 3 projects advancing liquid hydrogen jet engines. [cited 2023 23 August]. https://www.greencarcongress.com/2023/02/20230207-rr.html.

Green Car Congress (2024) Argonne, Achates Power developing hydrogen-powered opposed-piston engine. [cited 2024 8 June]. https://www.greencarcongress.com/2024/02/20240209-achates.html.

Grilli F et al (2020) Superconducting motors for aircraft propulsion: the Advanced Superconducting Motor Experimental Demonstrator project. J Phys: Conf Ser 1590:012051

H2FLY (2023) World's First Flight of Liquid Hydrogen Aircraft. 2023–09–07 [cited 2023 20 November]. https://www.h2fly.de/2023/09/07/h2fly-and-partners-complete-worlds-first-piloted-flight-of-liquid-hydrogen-powered-electric-aircraft/.

International Air Transport Association (2024) Aviation contrails and their climate effect: tackling uncertainties and enabling solutions

International Atomic Energy Agency (2018) Meet Oklo, the Earth's Two-billion-year-old only Known Natural Nuclear Reactor. 2018–08–10T12:45+02:00 [cited 2023 3 September]. https://www.iaea.org/newscenter/news/meet-oklo-the-earths-two-billion-year-old-only-known-natural-nuclear-reactor.

Loughran J (2023) Rolls-Royce nozzle breakthrough brings hydrogen plane engines closer to reality. Engineering and Technology Magazine. 2023-09-26T15:24:03+0000 [cited 2023 20 November]. https://eandt.theiet.org/2023/09/26/rolls-royce-nozzle-breakthrough-brings-hydrogen-plane-engines-closer-reality

Malayil J (2023) Aerospace firms join hands to further liquid hydrogen fuel systems. 2023–08–16 [cited 2023 23 August]. https://interestingengineering.com/transportation/aerospace-firms-join-hands-to-further-liquid-hydrogen-fuel-systems.

Mathias, P.M. and L.C. Brown (2003) Thermodynamics of the Sulfur-Iodine Cycle for Thermo-chemical Hydrogen Production. in 68th Annual Meeting of the Society of Chemical Engineers. Japan.

McCollom TM, Bach W (2009) Thermodynamic constraints on hydrogen generation during serpentinization of ultramafic rocks. Geochim Cosmochim Acta 73(3):856–875

Monolith Hydrogen (2021) Monolith receives conditional approval for one-billion dollar U.S. Department of Energy loan. [cited 2023 23 August]. https://monolith-corp.com/news/mon olith-receives-conditional-approval-for-a-one-billion-dollar-us-department-of-energy-loan? msclkid=8e0ef102d10a11eca7771e294270603b.

Monolith Hydrogen (2023) Process comparison. [cited 2023 23 August]. https://hydrogen.mon olith-corp.com/process-comparison.

Nature Portfolio (2023) Turbines driven purely by hydrogen in the pipeline. [cited 2023 23 August]. https://www.nature.com/articles/d42473-020-00545-7

Nica G (2016) Why did BMW really stop making the hydrogen 7 model? 2016-08-17 [cited 2022 25 September]. https://www.bmwblog.com/2016/08/17/bmw-stop-making-hydrogen-7-model/.

Nuttall WJ (2022) Nuclear Renaissance: Technologies and Policies for the Future of Nuclear Power (Second Edition). CRC Press.

Nuttall W, Glowacki BA, Clarke R (2005) A trip to 'Fusion Island'. 293:16–18

Nuttall, W.J., B.A. Glowacki, and S. Krishnamurthy (2016) Next Steps for Hydrogen. Institute of Physics.

Nuttall WJ, Glowacki BA (2010) Hydrogen as a fuel and as a coolant - from the superconductivity perspective. Journal of Energy Science. 1(1):15–28

Nuttall W, et al. (2020) Commercialising Fusion Energy - How small businesses are transforming big science, in Commercial opportunities for nuclear fusion.

Onuki K et al (2009) Thermochemical water-splitting cycle using iodine and sulfur. Energy Environ Sci 2(5):491–497

Payne J (2021) MIT to work on using tin to produce hydrogen without CO2 emissions. 2021–02–19 [cited 2023 23 August]. https://www.internationaltin.org/mit-to-work-on-using-tin-to-produce-hydrogen-without-co2-emissions/.

Prinzhofer A, Tahara Cissé CS, Diallo AB (2018) Discovery of a large accumulation of natural hydrogen in Bourakebougou (Mali). Int J Hydrogen Energy 43(42):19315–19326

Rolls Royce (2022) Rolls-Royce announces leading-edge hydrogen programme and developments in hybrid-electric research

RystadEnergy (2024) The white gold rush and the pursuit of natural hydrogen. [cited 2024 5 June]. https://www.rystadenergy.com/news/white-gold-rush-pursuit-natural-hydrogen.

Siemens Energy (2023) Zero Emission Hydrogen Turbine Center (ZEHTC). [cited 2023 23 August]. https://www.siemens-energy.com/global/en/home/products-services/solutions-use case/hydrogen/zehtc.html.

Sizewell C (2023) Hydrogen and SZC. [cited 2023 23 August]. https://www.sizewellc.com/enviro nment/szc-energy-hub/hydrogen/.

SuperTurbo Technologies (2023) Hydrogen boosting: a mechanically driven turbocharger system offering bidirectional power transfer and speed ratio control can improve air management in H2 engines. [cited 2023 14 August]. https://www.superturbo.net/media/articles/article-20/.

Team-BHP (2023) Largest hydrogen-powered passenger plane set to take flight. [cited 2023 22 August]. https://www.team-bhp.com/news/largest-hydrogen-powered-passenger-plane-set-take-flight.

The Guardian (2023) Prospectors hit the gas in the hunt for 'white hydrogen'. 2023-08-12 [cited 2023 22 August]. https://www.theguardian.com/environment/2023/aug/12/prospectors-hit-the-gas-in-the-hunt-for-white-hydrogen.

The Construction Index (2023) JCB has hailed as a breakthrough the conversion of a truck from diesel power to a hydrogen-fuelled combustion engine. https://www.theconstructionindex.co.uk/news/view/jcb-hails-hydrogen-engine-breakthrough

Universal Hydrogen (2023) Product. [cited 2022 29 September]. https://hydrogen.aero/product/.

Yazdani-Asrami M et al (2022) High temperature superconducting cables and their performance against short circuit faults: current development, challenges, solutions, and future trends. Supercond Sci Technol 35(8):083002

Young C (2021) Tiny 22-lb hydrogen engine may replace the traditional combustion engine 2021–05–21 [cited 2023 12 August]. https://interestingengineering.com/innovation/tiny-22-lb-hydrogen-engine-may-replace-the-traditional-combustion-engine.

Zgonnik V (2020) The occurrence and geoscience of natural hydrogen: A comprehensive review. Earth Sci Rev 203:103140

Chapter 6
The Way Ahead

Abstract This chapter concludes the book and points to some key ideas going forward. Rather than a future premised on the notion of colours of hydrogen the future of hydrogen is likely to be shaped by the need to manage two supply chains dedicated to separate products. One will be pipeline purity hydrogen and the other will be high purity hydrogen for mobile fuel cell applications. Ammonia is likely to be an important commodity, as may also be the case for other hydrogen carrier molecules. The world faces a pressing challenge to decarbonize and this book, and this chapter point to the fact that molecular energy carriers will have a major role to play. Electricity will grow greatly in importance in the low carbon future, but it is most unlikely to be the entire solution to the challenges that we face.

"Plans are worthless, but planning is everything"

President Dwight D. Eisenhower 14 November 1957 (The American Presidency Project 2023)

6.1 The End of the Rainbow?

In Chapter 1 we introduced the notion of the various colours allocated to different hydrogen production methods. Generally, we have argued that international public policy should only favour the production of low carbon hydrogen. Interesting and prominent examples of such hydrogen include: Green Hydrogen, Blue Hydrogen, Red Hydrogen and white hydrogen. In respect of Green Hydrogen we note, and include, negative emissions bio-energy with carbon capture and storage technologies (BECCS) such as those being proposed by Hydrogen Naturally (2023).

Hydrogen is hydrogen, whatever its provenance. If a framework of economic incentives and levies is established on the basis of the rainbow colours of hydrogen approach and applied to distribution and retail sales, then we see the risk of the following negative policy consequences:

© The Author(s) 2025   135
W. J. Nuttall et al., *Insights into the New Hydrogen Economy*,
https://doi.org/10.1007/978-3-031-71833-5_6

- Net-Zero compatible hydrogen risks not being produced by the lowest cost means (as best determined by market design and operation). A rainbow colours of hydrogen policy would be reminiscent of policy diktat and central planning with all the risks and inefficiencies usually associated with such approaches.
- In a system based on upstream provenance it will be burdensome to ensure certification of origin and there will be risk of illegal substitution and fraud.
- A system based on colours of hydrogen can act as a barrier to innovation in those areas disfavoured by policy. Policy aversion may be based on valid negative attributes of the disfavoured approach, but it could be that cost effective innovation would remove or avoid matters of concern. If policy had been poorly drafted, then it would be easy for new innovative ideas to be blocked by erroneous or outdated thinking. Indeed such realities risk steering inventors and entrepreneurs away from the low-carbon energy sector as a whole.

The consideration of environmental impacts from Green and Blue Hydrogen necessarily takes our thinking towards the distinction between the two methods of production. Despite the widespread discourse around Blue and Green Hydrogen and indeed around the rainbow of hydrogen colours presented in Chapter 1, the reality is that hydrogen is colourless and a molecule of hydrogen produced from renewable energy is identical to that produced during the partial oxidation of coal. Surely the central issue of concern is the environmental harm associated with the whole life-cycle of the activity and that should be assessed in operation, not prejudged based on the application of predictive labels applied generically.

The authors of this book look to a future where the notion of the colours of hydrogen is not a helpful mindset. A molecule of hydrogen carries with it no intrinsic attributes of sin or virtue. The issues of celebration or concern relate entirely to the processes adopted and not to the product itself. Within a trading community, such as that defined by the membership of the World Trade Organisation, or some other international arrangement, it should be possible to establish a policy regime that taxes or charges for the undesirable attributes of each and every hydrogen production process. Issues of concern include GHG emissions arising from system leaks. Prominent among such concerns is the issue of methane leakage associated with Blue Hydrogen production. We posit that once appropriate and project-specific charges are paid at the point of production, then the hydrogen generated can fairly be treated as equivalent with that produced by any other means. With the appropriate levies paid, all hydrogen should be treated equally. We propose this as a path to rapid greenhouse gas (GHG) emission reduction at scale.

The benefits of such an approach are several:

- There is no risk of illicit substitution of one colour of hydrogen with another – all hydrogen is equal in the retail market and hence there are no incentives to make such substitutions.
- There is no need to record provenance for hydrogen shipments with all the administrative burden that would entail.
- Policy-makers should be encouraged to apply low carbon charges based on the specific methods used in production and real emissions released. In this way

production innovations towards cleaner production, greater efficiency and lower costs can directly be incentivised.
* The wide and inevitably vague definitions hydrogen colours, which risk creating perverse incentives and distortions to innovation, are avoided entirely.

The virtue, or otherwise, of hydrogen lies in its ability to provide socially beneficial services with minimal negative environmental impacts. The virtue does not arise because the hydrogen product is branded as being 'green' or 'blue', what matters is the beneficial environmental impacts achieved and the meeting of energy needs. These developments are needed in a context where change is needed at scale and rapidly.

We note that the recently announced (in 2023) US Hydrogen Hubs span a range of hydrogen production methods. It appears that US policy is open to a range of 'colours' of hydrogen.

In the next section we will continue to take an optimistic view about the future of Blue hydrogen as we point a future of hydrogen no longer focussed on the various 'rainbow colours' associated with production methods. We said above that "*Hydrogen is hydrogen whatever its provenance*", but actually we do not take the view that there will simply be one type of hydrogen in the market, rather we are of the opinion that in practical terms there will be two.

6.2 Two Types of Hydrogen

We suggest that, in fact, there are two types of hydrogen – and the point of difference relates to the tolerated level of impurities. It is sometimes said that a Fuel Cell Electric Vehicle, based upon Proton Exchange Membrane (PEM) technology will run for years and years on good clean hydrogen, but that it will only run for only 20 s on dirty fuel. That is, we suggest that in the future hydrogen supply will fall into broadly two classes – pipeline purity and fuel cell purity. The latter will increasingly see a synergistic relationship to the cooling of superconducting magnetic technologies – a field known as 'hydrogen cryomagnetics', as discussed in Chapter 5.

Earlier in this book (for example the discussion of the UK in Chapter 4) we discussed the potential role for hydrogen as a new low carbon fuel for domestic heating and industry replacing today's pipeline natural gas. Nationally distributed pipeline hydrogen, even if based on newly prepared and lined pipeline infrastructures (which itself is an unlikely upgrade), almost certainly will never achieve the purity standard required for direct fuel cell use. There are also safety concerns that further render it unlikely that pipeline hydrogen could ever be used directly to fuel FCEV mobility. When odourless natural gas was introduced in the UK in the late 1960s replacing smelly (and toxic) town gas produced from the partial oxidation of coal, it was necessary to add an odorising agent (methanethiol, better known as methyl mercaptan) in very low concentration. The human nose is very sensitive, and the smell of odorised natural gas is quite distinctive. As with natural gas, the addition of an odorising agent would reduce the risk of a build-up of a hydrogen-air mix in a confined space (such as a dwelling or workplace). Even the small addition of a safety-related compounds, like methanethiol, could render hydrogen unsuitable for direct PEM fuel cell use (Wild et al. 2006). If pipeline hydrogen is to underpin a roll-out of PEM-FCEV fuel supply, for example via traditional filling stations now drawing their supplies from the former natural gas distribution grid, then significant clean-up capabilities would need to be provided in-situ on the forecourt. Such new infrastructures are likely to be technically challenging and expensive. Also, in the event of equipment failure and contaminated supply, with the consequent risk of serious damage to a customer's fuel cell, could render the whole proposition unattractive simply on the grounds of risk. It seems far more likely that, initially at least, hydrogen intended for FCEV use will be based on an independent high purity dedicated supply chain.

There is one approach which could perhaps simultaneously serve the needs of all hydrogen users – the cryogenic tanker distribution of liquid hydrogen (as discussed in Chapter 5). We posit that cryogenic liquid hydrogen might be attractive, despite the technical and hence economic challenges of liquefaction. The liquid could be available at sufficient scale to meet industrial and even heating needs, while the boil off gas from the liquid hydrogen, being the cleanest hydrogen of all, would be perfect for use directly in PEM FCEV applications.

If an international market can be established in which only ultra-low emissions hydrogen is traded, then the subsequent key issue for consumers will no longer be its mode of production nor its provenance but rather its purity. Fundamentally we suggest that there will be two types of hydrogen in the hydrogen economy. The first type will be very high purity hydrogen suitable for use in PEM FCEV and supplied by a dedicated industrial production and distribution system. The second type of hydrogen will be of somewhat lower purity and be odorised. It will serve the needs of home heating and industry, much as natural gas does today. Both types of hydrogen have the potential to be very large industrial activities, in the case of lower purity 'pipeline' hydrogen this is likely to be larger even than the whole electricity system today.

Green Hydrogen has much to contribute to the future of our planet, and its entry into the energy system at scale will initially be via developments in the electricity sector (including, but not limited to inter seasonal large-scale storage). These challenges within the future electricity system will be substantial but, in our opinion, they will fall short of the scale and immediacy of the Net-Zero challenge. The Net-Zero challenge means we need to be thinking on a much bigger scale than adjustments to our current electricity industry. One must not neglect the low-carbon benefits that will also come from the potential of a low carbon hydrogen (and synthetic fuel) industry emerging from today's polluting natural gas industry.

CCUS and Blue Hydrogen have the potential to be a force for good as the world faces a pressing need to decarbonise.

6.3 Hydrogen Must Have Low Emissions

It is important to note that there are at least three paramount factors for hydrogen's successful exploitation, as discussed in Chapter 1. There must be a **source** of hydrogen, there must be **infrastructure** by which hydrogen is supplied and there must be a **customer** willing to pay for hydrogen.

In the early 2020s there is an enormous wave of interest in hydrogen as a potential low carbon energy carrier. Attention is particularly devoted to two sources: Blue Hydrogen produced from fossil fuels accompanied by the implementation of CCS, and Green Hydrogen produced from renewable energy. This book tends to focus on Blue Hydrogen given the originating project's orientation to industry led activity. While there is a groundswell of industrial effort in many countries devoted to Green Hydrogen, much of the thought leadership around Green Hydrogen lies in academic and policy-making communities. In this project we have sought to give voice to industry and hence unsurprisingly Blue Hydrogen and related ideas have featured strongly. Concerning the possibilities for hydrogen infrastructure, these are dependent primarily on the scale of its supply, and on the distance over which it is to be supplied. In this book we have given attention to international supply chains in a global hydrogen economy rather than more local community-based approaches. There is a strong role for Green Hydrogen in the future energy system and indeed there will be examples of community initiatives in support of such a future. Our relative emphasis on top-down, national and international solutions should not be interpreted as any kind of criticism of other paths. We simply suggest that the processes emphasised in this book have the potential to be on a larger scale within a shorter time horizon.

Another key conclusion of this work is that all concerned with the future for hydrogen should have a good sense of the challenges and opportunities. Those coming to hydrogen from a background of renewable electricity generation can see hydrogen the potential future role of hydrogen as an inter-seasonal electricity storage option. They can take the view that it is potentially a large-scale activity, but in so doing they can miss a deeper truth – that the current electricity system in its entirety is not very big. On cold winter evenings in Great Britain the entire electrical energy supply is only approximately one sixth of total energy supply. Transport and mobility is 1/3rd (2/6ths) and heating is ½ (3/6ths). If the UK were to switch from oil to hydrogen for transport and mobility and from natural gas to hydrogen for heating then the scale of the future hydrogen economy would dwarf the entire electricity industry. Hydrogen has the potential to be far larger than mere summer to winter electricity storage. When deployed at a truly large scale, the colours of hydrogen will not be a matter of concern or interest. There will simply be a price for hydrogen as an unrestricted and fungible commodity. As Dan Sadler said at our first workshop in March 2021: "*you do not solve TW challenges with MW solutions*".

Another route to a similar insight concerning the potential scale of Green Hydrogen is to recognise that in early 2020s the UK uses roughly 2000 TWh of energy per year of which less than 10% comes from renewable sources. 2050 is only a few years away.

Thus far we have tended to explore the technological system but not the motivation for a change towards such systems. While the linked concerns of energy affordability and energy security are driving energy system change, especially since the renewed Russian invasion of Ukraine on 24 February 2022, the main motivation in favour of hydrogen as a new energy carrier and fuel is a reduction in harmful greenhouse gas emissions. The case for Green Hydrogen rests on the fact that normal operations of wind and solar photovoltaic energy have no harmful emissions. A full lifecycle assessment yields a more nuanced conclusion, see e.g. the manufacture of solar PV cells in China, but nevertheless renewable energy is undeniably a low-carbon alternative. The case for Blue Hydrogen is more contested. There are two aspects of concern. The first is that the sequestered CO_2 might leak to the atmosphere and hence contribute to harmful global warming. An even greater leakage cooling concerns methane the main component of natural gas and the main feedstock for the manufacture of Blue Hydrogen. The greenhouse gas emission footprints of both Green and Blue Hydrogen demand careful scrutiny and most be compared to the lifecycle impacts of electrification, for example, as the dominant strategic alternative. It is not the purpose of this project to provide detailed scientific estimates concerning greenhouse gas emissions impacts. Such issues were discussed in Chapter 3 and they represent a complex concern requiring expert scrutiny and careful policy consideration.

Throughout this book, and in respect of the points above, we have argued that ideally the emergent global hydrogen market should be developed without the need for provenance and trace-back. All hydrogen entering the market should be zero emission, or its producers must pay an internationally agreed rate to internationally agreed parties in respect of the environmental harms associated with its production. One such potential harm, however, is a particular concern for Blue Hydrogen developers – methane leakage. If this important aspect cannot be minimised in engineering terms and properly costed in economic terms, there is a risk that Blue Hydrogen will fail to live up to its full potential. In this book we see a bright and constructive future for Blue Hydrogen, but the methane leakage issue cannot be ignored. The issues of whole lifecycle emissions associated with Blue Hydrogen are complicated and somewhat contested (see e.g. (Clean Air Task Force 2021)). For those with an interest in the details of fugitive methane emissions and their role in the environmental case for Blue Hydrogen we recommend the work of researchers Bauer et al. (Bauer et al. 2022) and Antonini et al. (Antonini et al. 2020). In Fig. 6.1 we reproduce a figure from Bauer and co-workers. It shows the residual importance of fugitive greenhouse gas emissions for a range of Blue Hydrogen systems.

Figure 6.1 shows the importance of fugitive methane emissions reduction especially in scenarios of high CO_2 capture (93%). If Blue hydrogen can achieve an overall CH_4 emissions rate of 0.2% and 93% CO_2 CCS then the technology has the potential to be significant player in a low carbon future. It will not be a zero-emission energy source, but it would be a significant move to goodness. If however methane leakage is severe, then the technology could be environmentally worse than today's grey Hydrogen processes. Issues of hydrogen leakage have been discussed

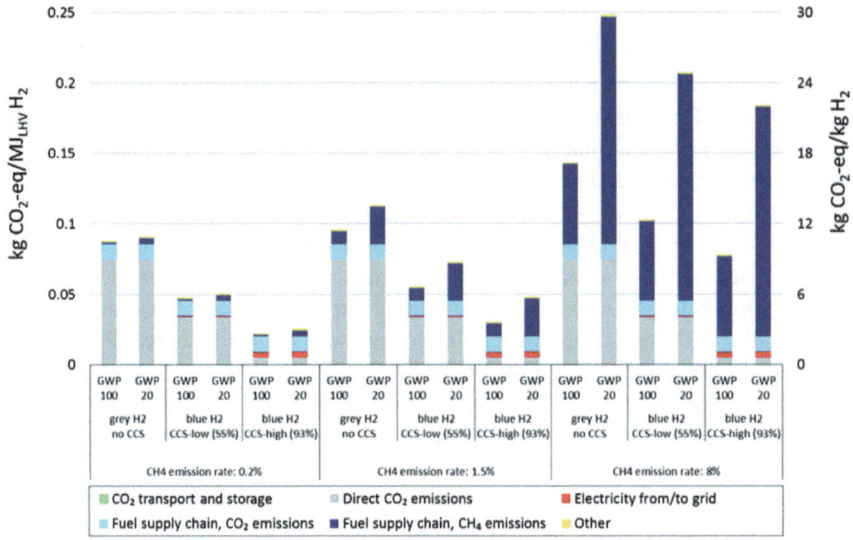

Fig. 6.1 The histogram illustrates life cycle emissions on 100 year and 20 year timescales associated with hydrogen production from natural gas (NG) in scenarios of varying methane emissions rates (0.2% - 8%) and overall CO_2 removal rates (55% and 93%) at the hydrogen production plant. GWP denotes Global Warming Potential where the subsequent number, 100 or 20, denotes the timescale of climate impact in years. Image rights: Reproduced from Ref. (Bauer et al. 2022) with permission from the Royal Society of Chemistry (CC-BY-3.0)

in section 2.2.5 and 3.1. Blue hydrogen needs to be a highly efficient and a very low leakage technology if it is to contribute in a positive way to a low carbon future. We believe the potential is there. There are two ways to seek to block poorly performing Blue Hydrogen operations one is legal and regulatory and the other is economic (by charging aggressively for all GHG emissions). While we would tend to prefer the latter approach for its technological neutrality, the more important thing is that poorly established Blue Hydrogen operations must be shut down either through the action of market forces (our preference) or direct regulatory intervention.

6.4 Ports and Pipelines

A key conclusion emerging from our project has been that while ports and pipelines are key to the future roll-out of hydrogen at scale, it will be the CO_2 pipelines that are the most important pipelines; indeed, they will be more important than future hydrogen pipelines. Such ideas resonated in both the online workshops held in 2021. A key part of that reality is that CO_2 is compressed easily into the liquid phase at ambient temperature (at around 5 MPa or 50 atmospheres – indeed with the triple point at 7.39 MPa and Tc of 31.04 °C a supercritical fluid state is also accessible). The ability to convey CO_2 as a liquid or dense fluid greatly simplifies pipeline solutions. Hydrogen has no such easily accessible liquid phases, although the research into hydrogen carrying liquid compounds, such as methyl cyclo-hexane, is interesting.

The Port of Rotterdam Authority suggests that a future hydrogen market might centre around ports and that this could be supported by the development of international shipping routes (International Energy Agency 2019). This is an interesting proposition for a number of reasons. Firstly, international shipping is itself a major emitter of CO_2 and could benefit from hydrogen as a means to decarbonise is power trains, as discussed in Chapter 3. Secondly, internationalisation could provide a means of de-risking hydrogen investments; should an expected market fail to materialise domestically, hydrogen could be shipped to overseas demand centres. Finally, the associated importation of CO_2 for Carbon Capture Utilisation and Storage (CCUS) could provide an additional source of revenue for hydrogen infrastructure centres if a suitable international CO_2 market oriented to globally optimised CCUS can be established.

When one considers the future of 'hydrogen clusters' one must pause to consider the synergies and complexities of at least three key molecules: methane (the dominant component of natural gas), carbon dioxide and hydrogen. One might also find oneself considering ammonia and potential hydrogen carrier molecules (such as methyl cyclohexane), but for simplicity let us consider just those three gases.

Let us first consider methane: there already exists a set of global clusters oriented to trading and storing this major energy commodity. Those existing infrastructures are likely to be extremely important in any shift towards the location of a cluster facilitating the end use of Blue Hydrogen.

As noted above, in the absence of existing infrastructure, CO_2 is usually more easily transported than hydrogen. Any country with an established natural gas cluster will however already have a natural gas distribution system. If it is straightforward to convert that to a hydrogen distribution system (see chapter 4 and the discussion of the UK) then it would seem likely that our natural gas hub could relatively easily transform into a true hydrogen hub. If, however, such pipeline hydrogen movements in existing networks is not possible for any reason, then it could be preferable to continue to pipe natural gas and manufacture hydrogen closer to the point of end use. In that case it would be possible to transport the CO_2 by tanker truck or low-cost pipeline for final CCUS, either back at the natural gas hub (if suitably located for CCUS) or to a third location if that proves preferable.

The interaction of considerations around natural gas origination, natural gas pipeline repurposing, sites for viable CCUS and sites of major hydrogen demand forms a relatively complicated optimisation problem for those contemplating a Blue Hydrogen cluster. Despite the challenges of building such an engineering system it appears that there are indeed locations in the world such as the Gulf Coast in the USA and Alberta Canada where rapid progress can be made. The Northeast of England, the Port of Rotterdam and Singapore may be other good examples.

Several existing natural gas clusters are associated with gas fields that are now depleting. As such it becomes necessary to replace a local geological resource with

Liquefied Natural Gas imports, hence the Blue Hydrogen energy clusters of the future are likely to involve Liquefied Natural Gas regasification.

In those places where the complicated considerations of a Blue Hydrogen cluster align in favour of such developments, the existence of abundant natural gas and CCS capabilities will favour the transition of existing industries (such as low carbon steel making) and the growth of new industries (such as long-term grid scale electricity storage) as part of the growing cluster. Such industries might also include synthetic fuel processing as an industrial successor to petroleum refining.

Not all linked activity need be geographically located within the cluster. Pipelines are a key part of the system to be considered. For example, a distant steel producer using blast furnace methods might take hydrogen from what had once been part of the natural gas transmission system. Alternatively, such an industry need not wait for low-cost hydrogen to become available at their facility. Such a steel producer could purchase natural gas based on existing distribution systems and operate its own natural gas reformer with associated CO_2 capture and compression. Supercritical or liquid CO_2 might then be trucked back to the main industrial cluster for environmentally responsible disposal.

When looking at the future for large-scale Blue Hydrogen roll-out, it seems likely that the future energy economy will be a natural gas economy, a hydrogen economy and indeed a carbon dioxide economy.

In imagining the roll-out of such a future it is important to appreciate that the investment will, and should, go to where the proposition is easiest. In the UK this coincides with a political desire to improve the parts of the UK that have arguably been left behind since 1980 as the economy shifted towards financial services and away from manufacturing.

Voices critical of Blue Hydrogen having a major role in the global push to reduce GHG emissions, rightly point to the need to reduce methane leakage. It must be remembered that the world is already operating a trillion-dollar natural gas economy with severe leakage issues. While of course ending that industry in its entirety could be said to eliminate those emissions that occur today. But that is not the only way to greatly reduce those emissions. Some Blue hydrogen solutions (in terms of clusters and networks) could greatly reduce methane leakage from today's high levels. We suggest it is erroneous to suggest that a shift to Blue Hydrogen would necessarily prompt an increase in GHG emissions compared to today's realities. Rather we suggest that the opposite is likely to be true and Blue Hydrogen is likely to be a major pragmatic opportunity in a global move to goodness. More research work on such ideas at the whole system level is required. That is not an easy task. In such work industry will be a source of much good insight as industry's existing concern for Scope 2 emissions is already focussing attention on leaks and other fugitive emissions which as we have discussed are key to any Blue Hydrogen proposition.

6.5   The Importance of Geography

A theme dating from the earliest stages of our discussions has been the importance of geography in determining the shape and scale of a future hydrogen economy and its industrial infrastructures. In particular we have considered the importance of clusters built around considerations of:

- Industrial heritage
- Access to feedstocks and resources
- Opportunities for carbon management
- Ports and pipelines access
- Supportive political environment
- Vibrant knowledge economy.

In the previous section we gave attention to the complexity of establishing Blue Hydrogen clusters in the context of long-standing natural gas capabilities. Hydrogen cluster opportunities are however not just restricted to Blue Hydrogen.

We observe that for places abundant in solar and wind resources but short of natural gas there will, of course, be an incentive to develop Green Hydrogen infrastructures, but we suggest that geopolitics and economics also matter. Another related consideration concerns regional (in the sense of continental) politics and policies. EU policy is one such example and indeed it has provided a significant impetus to Green Hydrogen in Greece and Spain. All states have renewable energy potential and the importance of this to building a low carbon energy system is obvious. Renewable energy resources also factor into the new geopolitics of energy security. Renewables have the benefit that the availability of commercial energy is not reliant on fuel supply chains, but typically renewable energy is reliant on seasonal, diurnal and weather-related factors. Nuclear energy is free of these concerns, but it is a relatively expensive technology and it brings with it its own special concerns (Nuttall 2022).

6.5.1   Key Markets for Hydrogen

The future geography of the hydrogen economy will not just be shaped by considerations of production, it will also be determined by the centres of demand. The success, or otherwise, of the hydrogen economy will be determined, in part, by the need to find customers. Hydrogen's early applications are likely to be specific to the locale of its market. Most studies (The Fuel Cell and Hydrogen Energy Association; The Hydrogen Council 2020; Fuel Cells and Hydrogen 2 Joint Undertaking 2019; E4tech 2016) suggest that one of the largest hydrogen customers at the initial stages will be in transportation, most specifically heavy-duty haulage, and mass transit in the form of trains and buses, as discussed in Chapter 3. Such customers can be expected to require relatively high purity fuel cell quality hydrogen (see above).

A key early focus for policy makers and infrastructure planners should be consideration of those places where transportation hubs coincide with the means of low-carbon hydrogen production and the infrastructure necessary to supply it. Such synergistic considerations can favour Green Hydrogen production methods with its greater ability to align with distributed generation and supply.

There is a need to identify markets where source, infrastructure and customer coincide. It is not simply enough to identify potential markets. As the hydrogen economy emerges in the coming decades major hydrogen developments will occur first in the locale where source, infrastructure and customer come together. As discussed in Chapter 4, there are already specific regions of the world with established hydrogen supply and demand supported by appropriate infrastructure. It makes sense to focus on these locations as potential areas for large-scale hydrogen industrialisation, but even then not to the exclusion of other viable smaller-scale, usually Green Hydrogen, options. Hydrogen production is expected to be an important area of innovation by many, as can be seen from the number of hydrogen-related roadmaps cited in this book. It is possible for a serendipitous convergence of capability and policy to capitalise on a hydrogen market emerging even before the necessary infrastructure is ready. Candidates for such a convergence should be identified and assessed alongside the established markets. Such locations will need co-ordinated strategy and planning across all aspects of supply and use including production, storage, distribution, supply and end use.

Whilst it is true that Green Hydrogen requires only electricity for its production, the nature of the local electricity supply will speak to the usefulness of Green Hydrogen to its chosen market. If, for example, the key local customer for hydrogen is an industry requiring pipeline quantities of hydrogen supplied continuously, then hydrogen produced by intermittent renewables must be linked to large-scale hydrogen storage. With that in mind, Blue Hydrogen might well be inappropriate for regions where large penetrations of intermittent renewable provide very low-cost electricity for significant periods. The price of electricity is, of course, more a consequence of electricity market design than it is of engineering costs. One key consideration in much of the developed world has been that grids must take renewable electricity first, when it is available. When renewable electricity supply exceeds demand then the electricity system operator will force curtailment of the renewable generation which can feel wasteful and motivate a desire to find alternative uses for the electricity generation, such as Green Hydrogen production. At such times the electricity price can be expected to be very low, or even negative, economically favouring Green Hydrogen ideas, but equally one can ask if something is indeed worthless what is the basis for one's sense of curtailment having been wasteful?

Our first event heard that the most important part of decarbonisation is to get an infrastructure developed able to accept both Blue and Green Hydrogen at scale. We heard expert opinion that Blue Hydrogen will dominate the transition, and once the infrastructure is in place, the Green Hydrogen will increasingly become more and more competitive and displace blue, over the decades to come. Earlier we suggested that in some key geographical locations Green Hydrogen will emerge first later facing competition from a global hydrogen industry based on the roll-out of Blue Hydrogen as a replacement for today's trillion-dollar natural gas economy. The notion that Blue Hydrogen will come first and gradually then be replaced by Green Hydrogen aligns with a logic that refers to Blue Hydrogen is a transition fuel, a role that some say natural gas is currently playing. We suggest that over the foreseeable future, let's say the next 100 years, both Green and Blue Hydrogen will emerge at scale. Their relative fortunes will ebb and flow over time in various locations each presenting evolving opportunities and challenges. Some territories will be dominated by Green Hydrogen while others will link much more strongly to Blue Hydrogen. Some regions will have only high purity fuel cell hydrogen while others will have infrastructures geared to lower purity hydrogen for industry and domestic heating. We posit, for example, that in the domain of pipeline hydrogen the UK has the potential to emerge as a world leading centre of excellence. Meanwhile we posit that global leadership in high purity fuel cell quality hydrogen for transport and mobility will emerge in North America.

We see signs that in North America the policy context is already sufficient for hydrogen to take off as a commercial proposition. Once sufficient incentives and subsidies are in place it can be the case that the most helpful thing that government can do is to get out of the way as industry seeks to do the heavy lifting of decarbonisation. Such a statement is, or course, far too simplistic because government has an ongoing role to ensure public and environmental safety and to help establish industrial norms and standards. One such standard setting role that is causing concern for the emergent hydrogen industry is the European Union's grid fees structure for hydrogen transmission (Simon 2023). Such a policy framework is crucial in order to incentivise investment in hydrogen pipeline infrastructures. European thinking is dominated by concern to find a space for Green Hydrogen and as things stand there is a gap between the production cost of Green Hydrogen and the price that European industry can afford to pay if such industries are to remain competitive in their respective global markets. In addition it is reported that Hydrogen pipelines extending over 20,000 kms are currently awaiting approval, but final investment decisions are on hold because of regulatory uncertainty (Simon 2023). A key uncertainty concerns the remuneration model applicable for new hydrogen transmission infrastructure. The issues at stake include the contentious and highly complex matter of cross border tariffs. It is feared that the resolution of such arcane aspects of energy policy may take ages to resolve and all the while hydrogen innovation remains stalled at a systemic level. Despite such cautionary comments, arguably the EU is showing the world the way – in that it is the first major market to grapple with key regulatory issues of the future. Despite the challenges faced by the world in the aftermath of the Russian invasion of Ukraine in 2022, the global energy system remains largely global. Regulatory concerns in one major market can have impacts in others. There is one recent 2023 story that perhaps better than any other reveals the tensions in play around local regulation and global competition (Kurmayer 2023). At the heart of the story is the still implicit assumption in Europe that hydrogen should be Green and that at some essential level it should align with the needs and interests of a growing electricity system centred on renewable generation. In such a scenario one needs electrolysers for hydrogen production using surplus electricity that are flexible in the face of rapidly varying levels of electricity generation and end use demand. Sometimes electricity will be cheap and plentiful and electrolysers should be able to ramp up fast to exploit such realities. In technical terms this takes one to PEM electrolysers and through its industrial policy and strategy the EU has developed a world-leading position in such technologies. It is not however the most inexpensive way to convert a stable source of electrical power into hydrogen for us. That accolade is held by a cheaper technology – the alkaline electrolyser. That is a technology with Chinese leadership. Both the EU and China are looking to the US market to underpin new globally oriented manufacturing ambitions. The question becomes what electrolyser standard will the US require, or will it even require a standard at all (Kurmayer 2023). These US regulatory questions risk having a determining role on EU industrial strategy. Will the US explicitly link its electrolyser requirements to the requirements of renewable electricity locking in a Green Hydrogen paradigm or will it adopt a more laissez-faire approach even at the risk of favouring Chinese players. As the time of writing in late 2023 the answer is not yet known but the sense that US is tending to favour technologically neutral decision making and that the rainbow colours approach is not of central importance going forward could be a sign of bad news ahead for the Europeans.

As noted earlier, natural gas is sometimes regarded as a "transition fuel". That logic implies that is the long-term natural gas has no sustainable future, but in the short to medium term it has a useful, albeit pragmatic, role to play in the global journey to a low-carbon future. We should declare that we do not see the roles played by various products and production methods being simply transitional. The factors are too complicated and locally determined for such a reality to be inevitable. Indeed, one can take the view that the era of natural gas could extend for longer than the era of gasoline. Should we regard petroleum as having been a "transition fuel"? We posit that such labels serve little real purpose and hence we do not present natural gas, and indeed Blue Hydrogen, in such terms.

Thus far we have suggested various factors such as geography, infrastructures and costs that will shape the emergence of the hydrogen economy. Another key consideration is economic risk. All projects will present intrinsic risks to be evaluated and considered. Even more importantly investors considering getting into the hydrogen business will need to assess external risks. Hydrogen will face numerous political and geopolitical challenges as it supersedes natural gas methane as the world's energy molecule of choice, as we posit it might. In addition to assessing the investors' expectations and risks, the broader economic, social, and environmental risks for the world should be considered.

6.6   Is Hydrogen in Trouble?

Generally we see rapid progress towards a new hydrogen economy with good progress in North America and the European Union. In its Global Hydrogen Review 2023 the International Energy Agency observes that:

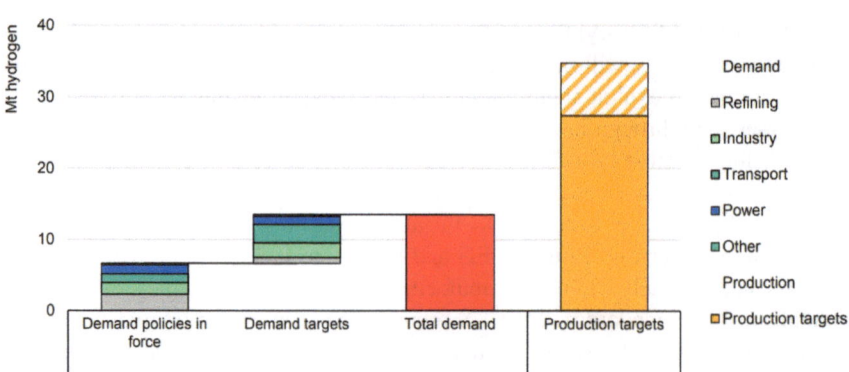

Fig. 6.2 Potential low-emission hydrogen demand in 2030 arising from implemented policies and government targets, and production targeted by governments. The IEA note that "in Production targets, the dashed area represents policy targets' ranges. For countries that do not have a production target, the low-emission hydrogen production is estimated from the capacity targets assuming a capacity factor of 57% and an energy efficiency of 69% for electrolysers and 90% for fossil-based technologies". (International Energy Agency 2023) Source: IEA CC-BY-4.0

> The number of announced projects for low-emission hydrogen production is rapidly expanding. Annual production of low-emission hydrogen could reach 38 Mt in 2030, if all announced projects are realised, although 17 Mt come from projects at early stages of development. The potential production by 2030 from announced projects to date is 50% larger than it was at the time of the release of the IEA's Global Hydrogen Review 2022.
>
> Taken from IEA's Global Hydrogen Review 2023 (International Energy Agency 2023).

The IEA's comment on production expansion appears to indicate that no difficulties lie ahead, but the IEA goes on to observe that growth in demand is not keeping pace with the growth in supply. This is illustrated in Fig. 6.2 (International Energy Agency 2023):

Importantly the IEA goes on to observe that:

> *the imbalance between demand and supply ambitions for low-emission hydrogen poses a risk for the announced hydrogen production projects to reach FID (only 4% have already done so) and, therefore, to achieve government targets for low-emission hydrogen production. This discrepancy could also impede the development of low-emission hydrogen supply chains.*

There is a need for hydrogen demand and supply to grow together and it is here that possible difficulties for effective hydrogen economy roll-out lie. If the lag in hydrogen demand growth is indeed a cause for concern, then we suggest that the attention of policy-makers and industry strategists should focus on both supply and demand simultaneously.

We observe newspaper headlines such as "Plans for Glasgow City Council to buy £7 million trucks SCRAPPED" from the Glasgow Times from 25 August 2023. The ending of such plans to use hydrogen point to problems in building up hydrogen demand (Hilley 2023). What are the origins of such demand growth difficulties? Interestingly, the newspaper appears to pin the blame for the bus company abandoning its plans on hydrogen supply. It reports that a council officer told the city's environment and liveable neighbourhoods policy committee that two contracts were cancelled by the designated hydrogen supplier in 2022 due to "programming difficulties". Furthermore, it would not be until 2027 there would be enough green hydrogen available to meet demand. Given the need to build demand growth it perhaps seems odd that the project was cancelled because of a local short-term shortage of hydrogen. The issue appears to have been a short-term lack of Green Hydrogen. A pragmatic view in favour of hydrogen more generally, and its role in building a low-carbon future, might have been more appropriate. Surely Grey Hydrogen could be sourced to run the trucks until low carbon hydrogen became available. It is important to understand that at any given time and place either hydrogen supply or demand will be in the lead. If one is waiting one should find a way to bridge the gap until the other half of the system can catch up.

Globally, at least, the IEA is saying that one can indeed expect new supplies to come on stream if demand is there. Whether the one news story from Glasgow is merely an anecdote or part of a wider story of "chicken and egg" difficulties around supply and demand for Green Hydrogen, we cannot say. We would say, however, as we have repeatedly in this book, that the fixation on the Rainbow Colours is generally not helpful to hydrogen roll-out or the vital goal of energy system decarbonisation. What is needed is for demand to have a technology neutral position on hydrogen supply chains as long as upstream policy properly takes account of the need to ensure dramatic whole system greenhouse gas emissions reductions. Some level of pragmatism during the transition is essential if success is to be achieved rapidly and efficiently.

The other area of demand requiring attention from policy-makers and industry strategists lies in the area of pipeline hydrogen for heating and industrial process uses. In the UK we see a great deal of mainstream media attention concerning the relative prospects of hydrogen boilers for domestic heating or the alternative use of electrical heat pumps. The majority of relevant stories in the media have been supportive of the heat pump proposition. For example on 21 October 2023 The Guardian newspaper in the UK ran a story under the headline: "Hydrogen boiler push to continue despite verdict of UK watchdog" (Harvey 2023). The watchdog referred to in The Guardian's headline is the National Infrastructure Commission. The report is based on the NIC's second national infrastructure assessment (National Infrastructure Commission 2023). The NIC's report recommends that the UK focus aggressively on energy system electrification. It recommends that hydrogen use is restricted to industry, surplus renewable electricity energy storage and power generation via combustion. It urges that hydrogen should not be used for domestic heating for which electric heat pumps are preferred. The report also calls for an end to unabated gas combustion for power recommending the addition of CCS to gas fired power generation. In contrast to the views of the NIC, however, we would suggest:

- CCS should be deployed first in those contexts where the best source of CO_2 is available. In that light we posit that Blue Hydrogen production is a far more attractive proposition than any CCGT-CCS power station investments. Flexible CCGT units are a useful component of an electrified future centred on renewables, but they are not the technology on which to first establish CCS at scale. Indeed SMR-based Blue Hydrogen production is a far better technology to link to CCS.

- We note UK consumer resistance to heat pumps on various grounds including high up-front cost (which the NIC recommends is mitigated through enhanced subsidy), poor heating performance in real use contexts (draughty old British housing stock – which might be improved through energy efficiency interventions), large space requirements (noting private outside space is often at a premium in the UK). We further note the greater consumer acceptance of straightforward boiler replacement moving towards 'hydrogen ready' technologies. Our main point is that the issues here are not primarily technology cost, but rather relate to the user experience and the social benefit of minimising lifestyle shift where possible. Costing is a tricky business and given the global nature of the decarbonisation challenge, cost assessment needs to be done at the whole system level irrespective of the subjective perspectives of the various actors (public and private) and where the funds come from (consumers bills or taxes).
- The need to use hydrogen for inter-seasonal electricity storage is advocated by the NIC largely as a response to a consequence of another of its recommendations – a very large increase in the use of intermittent and variable renewable electricity generation. In the absence of such electrification ambitions the role of hydrogen could be somewhat different. The level of inter-seasonality in UK power generation will be affected by UK policy for new baseload nuclear power plant construction. Such a large expansion is currently proposed (in 2023).
- The need for hydrogen in low carbon industry is recognised as being important at scale by the NIC. Even the NIC does not doubt that hydrogen is coming.

The NIC's 2023 presents an orthodox vision of green electrification. It is a vision that we in part challenge in this book. We note, as The Guardian reports, that the UK government remains somewhat unconvinced by the NIC's findings on hydrogen's future role. We rather share the UK government's view.

6.7 Molecules Matter

Repeatedly throughout this book we have referred to the global challenge of energy decarbonisation and the competition of ideas between electrification and hydrogen. As already noted, we are not suggesting that one will be at the expense of the other. Indeed, even as we stress the attributes of hydrogen, we acknowledge the benefits of electrification and expect a very substantial growth in the low carbon electricity sector to occur. Within hydrogen we have emphasised the issues of Blue Hydrogen and pointed to the alternative of Green Hydrogen. Similarly, despite drawing distinctions and even implying a level of competition, we expect both Green and Blue Hydrogen to grow significantly in importance. These two central comparisons – electrification versus hydrogen, and within hydrogen: Green versus Blue mask another important way to examine the core issues. The issue is not hydrogen – it is low carbon molecular fuels more generally. In chapter 2 we discussed the merits, or otherwise, of synthetic hydrocarbons (efuels), but the possibilities around future molecular fuels goes much wider than such ideas.

As we think about future molecular fuels, we realise that hydrogen will be much more than just a 'linkage agent' or an energy carrier – it will be an important part of a renaissance of chemical engineering for the energy economy. Chemical engineering will be a key part of the energy sector, belatedly, joining the knowledge economy which in recent decades has been dominated by the economics and policy of digitalisation. We suggest that low carbon molecular energy must join the knowledge economy. Chemical engineering has the potential, not to be in decline, but to be the enabler of global prosperity in a low carbon world.

One candidate future fuel that may hold potential especially for developing countries, and despite its status as a hydrocarbon, is "M85" methanol blend fuel (85% methanol and 15% gasoline), see Chapter 5, Box 5.6. Another interesting synthetic fuel is ammonia.

6.7.1 A Last Word on Ammonia

Looking beyond the ideas of Green Hydrogen, there is currently much interest in Green Ammonia. See for example the Haldor-Topsoe process for the production of Green Hydrogen via solid oxide electrolysis of water linked with exothermic ammonia synthesis. There are reports that the process is said to be 100% efficient, and hence 30% better than standard water electrolysis (Haldor Topsøe 2023). Ammonia is also attracting particular interest as a future maritime shipping fuel, see, for example: *Ammonfuel An industrial view of ammonia as a marine fuel (2020)* (Haldor Topsøe et al. 2020).

Molecules have several key benefits over electricity and electrification and one is global trade and this links to the minimisation of transmission losses. While there is talk of 1,000 km, or more, electricity transmission, such ambitions appear challenging in terms of infrastructure and inductive losses. Molecules, however, are much more easily shipped around the world and indeed any vaporisation can be used beneficially to fuel the ships involved. It is important therefore to stress the enduring importance of long-distance molecular energy vectors, as currently evidence by the global oil and natural gas industries. In future these industries must be shipping more environmentally benign energy molecules than the hydrocarbons of our fossil fuel economy (Smit and Powell 2023).

Ammonia has a potential role as a hydrogen carrier in international trade. The fundamental issue being that it is unfeasible for a global maritime trade economy to develop around gaseous hydrogen cargoes. The issue then becomes how to transport hydrogen in a condensed, i.e. liquid, state. Potential approaches include:

- Ammonia
- Methyl cyclohexane/Toluene

- Methanol
- Cryogenic Liquid Hydrogen
- Surface adsorbed hydrogen (zeolites etc.)

Of these various competing ideas ammonia is arguably already at the highest technology readiness level, as commercial ammonia cargoes are already shipped in support of the global fertiliser industry.

A key challenge to the deployment of ammonia as a hydrogen carrier is whether it will be compatible with the high purity requirements of PEM fuel cell technology. In that regard a recent announcement from AFC energy is particularly interesting. The company reports that their new ammonia cracker technology has produced 99.99% pure hydrogen from a single reactor test (UK Investor Magazine 2023). Such purity levels meet international standards for fuel cell applications.

UK high technology aerospace firm Reaction Engines has recently spun-out a subsidiary Sunborne Systems which has also recently reported significant progress in Ammonia cracking at hydrogen quality levels appropriate for PEM fuel cell use (Njovu 2023).

6.8   Hydrogen - the Way Ahead

We are grateful to Chris Johnson and Robin Little or their comments on the draft report prepared after the three events held in 2021 and 2022. They suggested that there are five live issues shaping the future of the hydrogen economy:

- *The effect of both regional and national incentives and penalties on the hydrogen production landscape.*
- *Public perception concerns (e.g. concerns around the future role of natural gas).*
- *How policy and regulation might be best shaped to nurture the emergence of large-scale low-emission hydrogen production.*
- *A need to assess the competitions and synergies between Green and Blue Hydrogen.*
- *A need for further thought on how the oil and gas sector might best strategically approach the development of Blue Hydrogen*

We agree with those suggestions and to that we would add five ways in our own thinking has evolved as a result of our HEIF funded events, and the development of this book.

- **We will continue to consider the idea of two types of hydrogen** We are inclined to be clearer in our mind when thinking about two types of hydrogen, defined by gas purity and not by production method, in the future hydrogen economy – odorised pipeline quality hydrogen for industrial and domestic heating purposes and high purity PEM fuel cell quality hydrogen for transport and mobility. We are interested in the notion that these products should be kept separate in distribution, but note the possibility, in principle, of clean-up from pipeline hydrogen to high purity PEM fuel cell hydrogen close to the point of retail sale.

- **Achieving Net-Zero across the entire global economy will require significant attention on CO_2 management and CCUS.** The importance of Carbon Capture, Utilisation and Storage (CCUS) will grow even in the absence of an emerging hydrogen economy, but developments in CCS have the potential to synergise efficiently with the emergence of Blue Hydrogen noting that such technologies are amongst some of the best sources of high concentration CO_2 for CCUS. Such ideas are discussed in Chapter 3, where it is further observed that it is generally easier to transport CO_2 than H_2. This reality further boosts the potential for Blue Hydrogen to play an important role in the global journey to Net-Zero in this century.

- **We accept that the success, or otherwise, of Blue Hydrogen will rely on close attention to natural gas leakage.** The natural gas industry is already a source of major methane leakage and a move to Blue Hydrogen could play a major role in reducing such emissions. The natural gas industry is already focussing on its scope 2 emissions, including such considerations. If leakages are not eliminated, then the case for Blue Hydrogen erodes rapidly. As noted in Chapter 3, the issue of indirect atmospheric impacts of emissions deserves further scrutiny.

- **We observe that Industrial clusters are key to Blue Hydrogen and clusters could also be very important for Green Hydrogen and hydrogen from nuclear energy.** In the case of Blue Hydrogen the issues are complicated involving legacy infrastructures associated with chemical engineering, geological resources, trade infrastructures and pipeline networks. It is important to note that unless established networks imply otherwise, it is generally easier to move CO_2 than hydrogen.

- **We continue to see the potential for powerful synergies between applied superconductivity and liquid hydrogen supply.** This idea has been termed 'hydrogen cryomagnetics' by Bartek Glowacki and co-workers (Nuttall and Glowacki 2010; Glowacki et al. 2015), and it is discussed in Chapter 5. We see the potential for these ideas to find their first major applications in low-emission aviation.

- **We will try to be as open-minded as possible as to current and future sources of low-emission hydrogen.** We note with interest the emergence of global awareness of white (or "gold") hydrogen from geological sources during our HEIF project and the subsequent development of this book (see Chapter 5 for insight into that story). We suggest that is a development meriting further attention. We also note that both Green and Blue Hydrogen are areas undergoing rapid innovation. In addition nuclear energy derived hydrogen (Red, Purple and Pink) is attracting much more attention, including in nuclear engineering circles (Nuttall 2022).

- **We will try to be as open-minded as possible as to current and future uses of low-emission hydrogen.** The emergence of a new hydrogen economy is likely to prompt the creation of whole new industries and to accelerate the transformation of existing enterprises.

As emphasised throughout this book hydrogen the potential to be a trillion-dollar industry. We have suggested that it has a role in energy far beyond simply balancing the future electricity system. We have further argued that hydrogen is much more than just an energy carrier. It is a potential future cryogen for high temperature supercon-ducting magnets in a wide range of applications, but also hydrogen is an important chemical for industrial use. One key role for hydrogen chemical engineering is as a reducing agent, for example in low-carbon steel production, as discussed in Chapter 3 The fact that hydrogen has chemistry is potentially hugely helpful to its transport as a cargo, see chapter 4 and to its role in enabling a range of low-carbon synthetic fuels. Of course, hydrogen has a direct role in transport and mobility, but this will likely not just to be a story of fuel cell electric vehicles. Internal combustion and gas turbine technology will arguably both have a significant presence in the low carbon future thanks to hydrogen.

One theme in the background throughout this project has been the role of inno-vation in the evolution of the hydrogen economy. It is important to distinguish inno-vation from invention. Of course, there is the potential for new inventions (such as the new Ammonia crackers mentioned earlier) but innovation is a wider and even more powerful concept. Fundamentally it concerns new ways of doing things. There will be innovation within hydrogen as production, distribution and use are all improved through incremental adjustments and through more radical jumps. But the real power of hydrogen is the way in which it permits innovation in whole sectors of the economy such as steel making, energy storage, domestic heating and mobility. The emergence of low-carbon aviation appears to be a particularly rich and powerful area of innovation and as discussed in Chapter 5 hydrogen could play a major role there. In essence all areas of innovation represent opportunities for business change and business growth. Agility is likely to be a powerful business attribute as hydrogen moves forward. Start-up companies are famously agile and risk-taking and they could be particularly important in revealing the opportunities in the emerging hydrogen economy.

The world of start-up innovation is dominated in the 2020s by the experience gained over the last 40 years in the global digital economy. High technology digital enterprises completely overtook the communications world. Companies such as Microsoft, Apple and Meta overtook IBM, AT&T and Motorola in terms of stock valuation and public awareness. Will a similar phenomenon occur in energy innova-tion? What will the role of energy incumbents be? These are not easy questions and we certainly do not know the answer. All we can say is do not assume that hydrogen innovation will look like digital innovation. It could be that incumbents evolve and thrive based on a high level of expert insight and understanding. Arguably energy requires a larger innovation unit and a different relationship to skills and capital than an internet start-up. Having said that three gradates on the science park might have an idea that becomes a billion-dollar hydrogen business – we are not saying it can't happen.

We close with an historical analogy, conscious of the risks and dangers of such comparisons. It can be argued that the nineteenth century was the age of steam (building upon eighteenth century science and engineering ideas). It can further be argued that the twentieth century was the era of electricity (building upon nineteenth century science and engineering). We might posit that the twenty-first century could be the age of hydrogen (building upon twentieth century science and engineering). Such an analogy allows one to make some follow-on observations:

- The use of electricity will expand during most of the twenty-first century, just as the use of steam expanded through much of the twentieth century.
- In the 2020s the electricity industry will rightly secure much capital investment and attract much human capital as it seeks to roll out its benefits to society more widely and deeply. This is analogous to the role of steam locomotive design and development in the 1920s. In the early Twentieth Century one might have imagined that steam power would decline, as other technologies (e.g. diesel locomotives) appeared. In reality the world's best steam locomotive engineering was still to come. The emergence of competing technologies (including superior technologies) even boosted steam innovation, before the inevitable decline.
- Electricity did not triumph in the twentieth century because it was the cheapest way to light a city at night or to drive factory machinery by day, it succeeded because it aligned with user needs (electricity was clean, convenient and increasingly reliable) and it also sat well with the Zeitgeist of the 1930s and modernism. Similarly, hydrogen transport and mobility sells itself not on price, but on other attributes. While hydrogen and clean syn-fuels triumph in the Twenty-First Century? We cannot say.

We can posit, however, that the emergence of a new clean molecules energy economy, based on hydrogen, might spur the development of whole new ancillary technologies. One such technology to posit is hydrogen cryomagnetics involving practical superconductors (see Chapter 5, Box 5.4). As Bartek Glowacki from the University of Cambridge (who has developed many practical low and high temperature superconductors) has observed, when high temperature superconductivity (HTS) emerged in the 1980s it was widely assumed that liquid nitrogen would be the key to its success. As it appears 20 years later that the electromagnetic performance of HTS superconductors at liquid nitrogen was not satisfactory and $MgB2$ practical superconductor could not be cooled by liquid nitrogen, he predicted that the key cryogen for HTS will be hydrogen not nitrogen. Today 40 years after the discovery of HTS, the emerging hydrogen economy presents a real opportunity for hydrogen cryomagnetic applications.

In this book we posit a bright future for the hydrogen economy. We deliberately take an optimistic line concerning all aspects of hydrogen innovation. The twenty-first century is not yet the hydrogen century, and we concede it may never be so, but in this book we hope he have been able to provide some insights beyond the basics to explain why we think that hydrogen has real potential to deliver beneficial change at scale by 2050 and thereafter it has the potential to define much of the energy system of the late twenty-first century. Electrification is growing and will continue to grow fast, but hydrogen (and its associated syn-fuels) will, we suggest, one day be bigger.

References

Antonini C et al (2020) Hydrogen production from natural gas and biomethane with carbon capture and storage—a techno-environmental analysis. Sustain Energy Fuels 4(6):2967–2986

Bauer C et al (2022) On the climate impacts of blue hydrogen production. Sustain Energy Fuels 6(1):66–75

Clean Air Task Force (2021) Putting the Howarth & Jacobson hydrogen paper in context: decarbonization and the potential greenhouse gas emissions performance of "blue" hydrogen

de Wild PJ et al (2006) Removal of sulphur-containing odorants from fuel gases for fuel cell-based combined heat and power applications. J Power Sources 159(2):995–1004

E4tech (2016) Hydrogen and fuel cells: opportunities for growth—a roadmap for the UK

Fuel Cells and Hydrogen 2 Joint Undertaking (2019) Hydrogen Roadmap Europe

Glowacki BA et al (2015) Hydrogen cryomagnetics for decentralised energy management and superconductivity. J Superconduct Novel Magn 28(2):561–571

Haldor Topsøe (2023) Green ammonia. [cited 2023 12 August] . https://www.topsoe.com/proces ses/green-ammonia

Haldor Topsøe et al (2020) Ammonfuel—an industrial view of ammonia as a marine fuel

Harvey F (2023) Hydrogen boiler push to continue despite verdict of UK watchdog. 2023–10–21 [cited 2023 30 October]. https://www.theguardian.com/environment/2023/oct/21/hydrogen-boi ler-home-heating-uk

Hilley S (2023) Plans for Glasgow City Council to buy £7 million trucks SCRAPPED. [cited 2023 30 October]. https://www.glasgowtimes.co.uk/news/scottish-news/23748519.plans-glasgow-city-council-buy-7-million-trucks-scrapped/

Hydrogen Naturally (2023) Bright Green™ Hydrogen. [cited 2023 15 September] . https://www.h2naturally.com

International Energy Agency (2019) The future of hydrogen

International Energy Agency (2023) Global Hydrogen Review 2023. Paris

Kurmayer NJ (2023) US hydrogen rules could decide fate of EU's electrolyser industry. 2023-09-08 [cited 2023 30 October]. https://www.euractiv.com/section/energy-environment/news/how-us-hydrogen-rules-could-decide-fate-of-eus-electrolyser-industry/

National Infrastructure Commission (2023) Second National Infrastructure Assessment

Njovu G (2023) Sunborne Systems & AFC Energy: successful ammonia cracker demonstrations [cited 2023 30 October]

Nuttall WJ (2022) Nuclear renaissance: technologies and policies for the future of nuclear power, 2nd edn. CRC Press, Boca Raton

Nuttall WJ, Glowacki BA (2010) Hydrogen as a fuel and as a coolant - from the superconductivity perspective. J Energy Sci 1(1):15–28

Simon F (2023) Grid fees remain key sticking point as EU finalises hydrogen rules. 2023-09-28 [cited 2023 30 October] . https://www.euractiv.com/section/energy-environment/news/grid-fees-remain-key-sticking-point-as-eu-finalises-hydrogen-rules/

Smit DJ, Powell JB (2023) Role of international oil companies in the net-zero emission energy transition. Annu Rev Chem Biomol Eng 14(1):301–322

The American Presidency Project. Remarks at the National Defense Executive Reserve Conference (2023) [cited 2023 15 September]. https://www.presidency.ucsb.edu/documents/remarks-the-national-defense-executive-reserve-conference

The Fuel Cell and Hydrogen Energy Association, Road Map to a US Hydrogen Economy

The Hydrogen Council (2020) Path to hydrogen competitiveness: a cost perspective

UK Investor Magazine (2023) AFC Energy shares rise on ammonia cracking hydrogen test. 2023–10–23 [cited 2023 30 October]. https://ukinvestormagazine.co.uk/afc-energy-shares-rise-on-ammonia-cracking-hydrogen-test/

Index

© The Editor(s) (if applicable) and The Author(s) 2025
W. J. Nuttall et al., *Insights into the New Hydrogen Economy*,
https://doi.org/10.1007/978-3-031-71833-5